Lecture Notes in Networks and Systems

Volume 207

The series "Lecture Notes in Networks and Systems" publishes the latest developments in Networks and Systems—quickly, informally and with high quality. Original research reported in proceedings and post-proceedings represents the core of LNNS.

Volumes published in LNNS embrace all aspects and subfields of, as well as new challenges in, Networks and Systems.

The series contains proceedings and edited volumes in systems and networks, spanning the areas of Cyber-Physical Systems, Autonomous Systems, Sensor Networks, Control Systems, Energy Systems, Automotive Systems, Biological Systems, Vehicular Networking and Connected Vehicles, Aerospace Systems, Automation, Manufacturing, Smart Grids, Nonlinear Systems, Power Systems, Robotics, Social Systems, Economic Systems and other. Of particular value to both the contributors and the readership are the short publication timeframe and the world-wide distribution and exposure which enable both a wide and rapid dissemination of research output.

The series covers the theory, applications, and perspectives on the state of the art and future developments relevant to systems and networks, decision making, control, complex processes and related areas, as embedded in the fields of interdisciplinary and applied sciences, engineering, computer science, physics, economics, social, and life sciences, as well as the paradigms and methodologies behind them.

Indexed by SCOPUS, INSPEC, WTI Frankfurt eG, zbMATH, SCImago.

All books published in the series are submitted for consideration in Web of Science.

More information about this series at http://www.springer.com/series/15179

Elżbieta Macioszek · Grzegorz Sierpiński
Editors

Research Methods in Modern Urban Transportation Systems and Networks

Springer

Editors
Elżbieta Macioszek
Faculty of Transport and Aviation Engineering
Silesian University of Technology
Gliwice, Poland

Grzegorz Sierpiński
Faculty of Transport and Aviation Engineering
Silesian University of Technology
Gliwice, Poland

ISSN 2367-3370 ISSN 2367-3389 (electronic)
Lecture Notes in Networks and Systems
ISBN 978-3-030-71707-0 ISBN 978-3-030-71708-7 (eBook)
https://doi.org/10.1007/978-3-030-71708-7

This Springer imprint is published by the registered company Springer Nature Switzerland AG
The registered company address is: Gewerbestrasse 11, 6330 Cham, Switzerland

Preface

The transport system of every territory is a complex set of technical sub-systems: branched, functional, organisational, financial, and regulatory. Its backbone is a network of connections which determines the transport accessibility of the given territory. Over the recent years, the sphere of transport has been systematised to a large extent, and yet the existing transport systems remain not entirely efficient, sometimes obsolete and incomplete. What seems to be necessary on account of this fact is acting consistently towards improvement of transport networks and systems, but in the first place, research aimed at identification of the problems at hand as well as their elimination by means of adequate methods. Moreover, the contemporary dynamic development of various areas of human activity is the origin of further new goals that cities must pursue, often facing consequences of the significant nuisance caused by the growing road traffic volumes. When appropriately used, the available array of tools and methods may contribute to alleviating the deleterious impact of high traffic volumes on the traffic conditions in cities. Thanks to well-matched research methods, urban transport networks and systems can develop, thus providing the foundation for numerous transport sectors such as road transport, collective public transport, or rail transport. With regard to the foregoing, the main goal of the monograph is to demonstrate and discuss the issues related to the research methods and solutions currently used in urban transport networks and systems. This monograph is a detailed treaty on such specific subjects as the research methods and solutions used in urban transport networks and systems, with special emphasis on the methods used in the spheres of urban safety and urban behaviour, as well as eco-friendly solutions. With regard to the environmentally friendly solutions, the publication elaborates on such methods as, for instance, those which take air quality into account while considering the impact of the atmospheric emission from urban road traffic, or the Maja multi-criteria method for assessment of operation of vehicles with different power transmission systems in the urban environment, a life-cycle costing decision-making methodology, as well as an urban intersection design method which enables modelling and analysis for a circular city. Special attention has been paid to the research methods related to urban safety and human behaviour in the urban environment, since they determine, both indirectly and directly, the safety level in urban transport networks and systems, and primarily address the problems of human health and life. These include modelling of vehicle speed in city streets by taking the driver's behaviour into account, a method of assessing the conditions of the bike-sharing system operation and its implementation in medium-sized cities and speed-related surrogate measures of road safety based on floating car data, as well as a global approach based on the application of the method of cross-sectional surveys in studying parents' attitude towards safe transport of children in Europe.

The presented monograph, entitled "Research Methods in Modern Urban Transportation Systems and Networks" provides an excellent opportunity to learn about the latest trends and achievements in the field of research methods in modern urban transportation systems and networks. We would like to take this opportunity to kindly invite you to read the monograph. Readers interested in current achievements in the field of research methods in modern urban transportation systems and networks, problems and directions of development of transport and traffic engineering will find in the monograph extensive material presenting the results of scientific research, various views and considerations, as well as new approaches and methods of solving problems. In connection with the above, we wish all a fruitful reading.

Katowice, Poland
January 2021

Elżbieta Macioszek
Grzegorz Sierpiński

Contents

Chapter 1
Method of Ensuring the Technical Readiness of Transport Companies Fleet Due to the Region's Capabilities

Serhii Holovnia, Vitalii Naumov, and Olha Shulika

1.1 Introduction

Contemporary motor transport companies usually include multi-brand vehicles. To ensure the technical readiness of the fleet of various brands, it is necessary to have the appropriate vehicles for each brand, trained repair specialists, stocks of spare parts, etc. It requires the involvement of significant material and human resources. Therefore, as a variant, the organization of maintenance and repair (M&R) works of motor vehicles can be considered when not only possibilities of own enterprises, but also various motor service companies are involved for the performance of works. However, when attracting third-party motor service companies, including them in the system of M&R of the motor transport company (MTC), it is necessary to solve the problem of an appropriate choice of a motor service company (MSC) in the region among their existing diversity. The task of the choice is to consider not only the economic feasibility of the transfer of M&R measures to the MSC but also to ensure the technical readiness of the fleet. At the same time, the technical readiness is affected by the remoteness of the MSC from the vehicles, the duration and probability of service by each MSC, the companies' own workload with clients, etc.

The problem of ensuring the level of technical readiness of MTC fleet due to the region capabilities can be formulated as follows. For each k-th MTC ($k =$

S. Holovnia
National Academy of the State Border Guard Service of Ukraine, Str. Shevchenka 46, 29000 Khmelnytskyi, Ukraine

V. Naumov (✉)
Cracow University of Technology, Str. Warszawska 24, 31155 Kraków, Poland
e-mail: vnaumov@pk.edu.pl

O. Shulika
Kharkiv National Automobile and Highway University, Str. Yaroslava Mydroho 25, 61002 Kharkiv, Ukraine

E. Macioszek and G. Sierpiński (eds.), *Research Methods in Modern Urban Transportation Systems and Networks*, Lecture Notes in Networks and Systems 207, https://doi.org/10.1007/978-3-030-71708-7_1

$1, 2, \ldots, N$) in the region, there are $\left(n_1^k, n_2^k, \ldots, n_M^k\right)$ vehicles of M brands that can be represented by the vector $\mathbf{N_Q}$:

$$\mathbf{N_Q} = \left(n_1^1, \ldots, n_M^1, \ldots, n_1^N, \ldots, n_M^N\right) = \left(n_1^1, \ldots, n_Q^N\right). \tag{1.1}$$

Annual volume Q ($Q = N \cdot M$, where Q is the required number of recipients involved in the distribution of services) of b_j^k M&R measures should be evaluated in advance and then distributed between Z servicing companies (A_1, $A_2, \ldots,$ A_Z) that may include repair stations of transport companies. After grouping the vehicles of each k-th MTC by M brands, the sources of requests for M&R can be represented as $\left(B_1,\ B_2, \ldots,\ B_Q\right)$. Each of them generates $\left(b_1,\ b_2, \ldots,\ b_Q\right)$ requests for M&R per year. Third-party MSCs and repair units of the transport companies (A_1, $A_2, \ldots,$ A_Z) have the corresponding annual production capacities (a_1, $a_2, \ldots,$ a_Z), i.e. the number of M&R services that can be performed per year. Since it is important for clients not to waste time waiting in queues and not to be rejected due to the overload of MSCs, the values (a_1, $a_2, \ldots,$ a_Z) should be determined taking into account the permitted probabilities for the service operations.

It's asserted that the following indicators are known: the distances from each MTC to each MSC, costs of each type of service, the number of service channels (servicing mechanisms), productivity, and the parameters of demand for services. The time interval between the requests for M&R services and the duration of maintenance may be considered as stochastic variables.

For each source of requests $\left(B_1,\ B_2, \ldots,\ B_Q\right)$, such the requests distribution $\left(b_1,\ b_2, \ldots,\ b_Q\right)$ should be obtained for (A_1, $A_2, \ldots,$ A_Z) organizations according to their annual production capacities (a_1, $a_2, \ldots,$ a_Z) that considers the existing workload of the servicing companies (δ_1, $\delta_2, \ldots,$ δ_Z). Based on the obtained distribution, the downtime of servicing mechanisms should be minimized, and the corresponding servicing costs should be optimized.

1.2 Literature Review

The issues of choice and distribution of various resources were considered in numerous publications. The following problems are usually solved by researchers: the optimal distribution of the general plan of the industry by enterprises; the optimal distribution of capital investments; the distribution of vehicles by flows; the distribution of employees by wages; the distribution of the population by income, etc.

The application of multi-purpose optimization problems in maintenance planning is considered in the paper [1], where reliability, availability, maintainability, and cost are used as decision criteria in the model. The authors state that the appropriate development of each service strategy depends not only on service intervals but also on the resources (human and material) that available for the implementation of such strategies. In the paper [2], an integrated model of optimization of a comprehensive maintenance plan based on a supply chain of services in high demand, considering the possibility of involving third-party organizations in the service. The authors analyze the allocation of resources in accordance with several requirements from the point of view of service providers; investigate the method of reasonable allocation of service staff and devices used by outsourcing companies for maintenance. The authors of the publication [3] also investigated the distribution of resources in conditions of their limitations. However, the focus of the research is to address the issue of stabilization of the basic schedule of the project in the presence of changes in the duration of activities.

On the other hand, if we analyze the work related to planning, organizing, and optimizing the maintenance of equipment and technical devices in general, they can be divided into several groups: reliability assessment by deterioration modeling [4, 5], comparisons of different maintenance alternatives [6, 7], and multi-objective optimization techniques [4, 5, 8, 9].

The first group of works mainly consists of research on ways to improve equipment reliability through condition-based maintenance. At the same time, researchers are developing strategies for maintenance that collects and assesses real-time information and recommends maintenance decisions based on the current condition of the system using computer-based monitoring technologies.

In the second group of studies that are based on the choice of different maintenance alternatives authors, mainly, consider the solution of the extreme value problem of objective function under linear constraints. In this case, deterministic and stochastic planning models are proposed by the researchers [6, 7, 10–14].

Papers of the third group are distinguished by the most modern and comprehensive approach. Search-based optimization techniques are used for maintenance planning considering the multiple objective functions and resource constraints. One of the simplest methods are approaches based on genetic algorithms [4, 5]. In the publication [5], by applying the Markov chain process in conjunction with genetic algorithms the authors solve the problem of forecasting the bridge's performance. In the paper [8], a multi-scenario optimization model with the method of scenario planning was used to solve the reusable logistics resource allocation problem. Within developing tactical production plans in a multi-stage production system, the papers [6, 9] propose to use a deterministic planning model that integrates the variability of the finished-product demands.

To sum up, it is necessary to emphasize that a search-based optimization solution addresses multiple maintenance planning objective but does not consider the subjective preferences of asset owners or decision-makers. There are no known models for solving problems that would comprehensively consider these conditions, as they usually consider only deterministic stocks and cargo needs based on the minimum cost or duration of the transportation plan. Although stochastic models have received more attention recently, existing stochastic models are still very simplified and lack a holistic approach. Another significant problem is the conversion of the transport companies vehicle fleet to electric vehicles [15, 16].

1.3 Proposed Methodology

To assess the effectiveness of the distribution results, it is proposed to use the following indicators:

- KD_{ij}—the coefficient of downtime of the *j*-th vehicle under the condition of servicing by the *i*-th MSC,
- C_{ij}—the maintenance costs of the *j*-th unit at the *i*-th MSC,
- KDC_{ij}—the complex indicator that is the product of KD_{ij} and C_{ij}.

The indicator KD_{ij} assesses the technical readiness of the *j*-th vehicle during maintenance at the *i*-th MSC or the MTC repair unit. The indicator is calculated as the ratio of the sum of the servicing time T_S and the relocation time T_R to the considered time period T_P:

$$KD = \frac{T_S + T_R}{T_P}. \tag{1.2}$$

We propose to estimate the value T_S considering the waiting time of vehicles in the MSC queue and the possibility of receiving of group requests [17]:

$$T_S = \frac{q}{\mu} + \sum_{k=0}^{w-1} P_{n+k} \cdot \frac{k+1}{n \cdot \mu}, \tag{1.3}$$

where:

P_{n+k} the probability of a state when *n* service channels are occupied and *k* requests are in the queue ($k = 1 \ldots w$, *w* is the permitted size of the queue),
n the number of service channels (M&R stands) of MSC or MTC repair unit,
q relative capacity of the servicing system,
μ service intensity of a channel

The total servicing costs *C* include the cost of M&R services C_{MR} and the cost of moving the vehicle to the servicing station and back to its permanent location C_T:

$$C = C_{MP} + C_T. \tag{1.4}$$

The travel costs C_T for maintenance at the MSC located at the distance *d* is calculated by the formula:

$$C_T = (h_{100} \cdot + d_{100}) \cdot \frac{2 \cdot d}{100} \cdot H, \tag{1.5}$$

where:

h_{100} the fuel consumption for the relocated vehicle,

c_f the fuel cost for the vehicle being relocated for M&R,
d_{100} depreciation costs per vehicle covering distance of 100 km,
H the number of movements to the MSC or MTC repair unit for maintenance of the vehicle during the considered period of time T_P

A complex indicator KDC is the efficiency criterion for a given MSC or the MTC repair unit. It characterizes the advantages (disadvantages) of the assessed servicing unit in comparison with similar companies (enterprises) that provide maintenance services.

We propose to solve the problem by using the sequence of stages shown in Fig. 1.1.

The initial data for the calculation in the form of matrices and vectors are presented by the vector **B** of requests for services, the vector **A** of production capabilities of MSCs, the matrices $\mathbf{T_S}$ and $\mathbf{T_R}$ of service times and relocation times, the matrix of downtime coefficients **KD**, and the matrix **C** of maintenance and relocation costs.

The vector of requests **B** contains the values of b_{ij} (the number of requests for maintenance of the j-th vehicles at the i-th MSC) calculated as a sum of the number of maintenance services $y_M^{u,r,k}$ and repair services $y_R^{u,r,k}$ needed for the u-th type ($u = 1, 2, \ldots, N_T$, N_T is the number of considered vehicle types) of vehicles of the

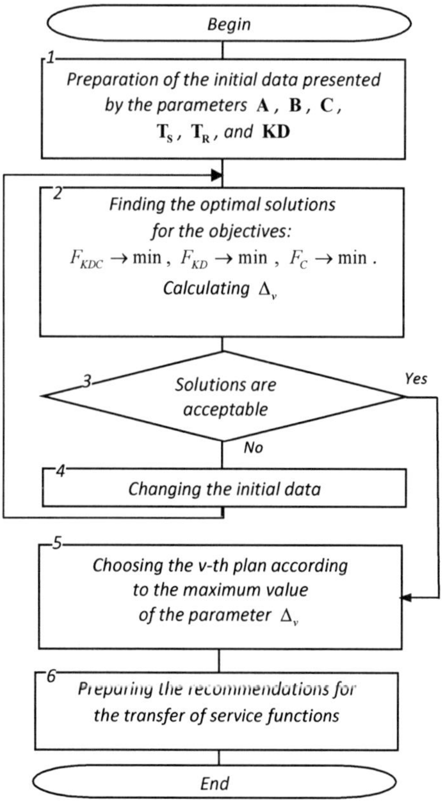

Fig. 1.1 The method of ensuring the technical readiness of companies' fleet

r-th brand at the region of the k-th MTC during the year. Values $y_R^{u,r,k}$ are calculated based on the regression models representing dependence of the number of repairs on the service lifecycle for the given type of vehicles. If the vector **B** contains zero values, these elements can be removed from the vector, because they will not be considered in the problem's solution.

The values of the vector **A** elements representing the production capacities of servicing companies are determined based on the estimated intensities I_i of the flow of M&R requests and the intensities δ_i of the flows of the MSC's own customers:

$$a_i = (I_i - \delta_i) \cdot T_P, \; i = 1 \dots Z. \tag{1.6}$$

The values I_i are defined in accordance with the probability $P_{s(i)}$ of servicing the requests by the i-th MSC. The value $P_{s(i)}$ must be not smaller than the given allowable probability P_S of service: $P_{s(i)} \geq P_S$.

To calculate the intensity I, we propose to use the following expression:

$$I = \frac{\mu}{P_S} \cdot \left(n - \sum_{k=0}^{n} (n-k) \cdot P_k \right), \tag{1.7}$$

where:

P_k the probability of the system being in the state when k ($k = 1 \dots n$) service channels are occupied

The matrix **KD** is the result of estimating the average value of the downtime coefficients during the maintenance by each of the servicing companies. Based on the **KD** and **C** matrices, the matrix **KDC** is calculated. If the MSC is unable to maintain a model or the given vehicle, the positive infinity value is assigned to the corresponding element of the **KDC** matrix.

The decision variables matrix $\mathbf{X_{ZQ}}$ represents the distribution plan and is based on the data from matrices **A**, **B**, **C**, and **KD**. Each of Z rows of the matrix $\mathbf{X_{ZQ}}$ correspond to the i-th servicing company considered in the problem. Each of Q columns of the matrix correspond to the number of vehicles of a brand belonging to the k-th MTC.

To solve the vehicles' distribution problem, based on completed **KD**, **C**, and **KDC** matrices, it is necessary to find such non-negative values of the $\mathbf{X_{ZQ}}$ matrix elements that satisfy the following system of linear equations:

$$\begin{cases} \sum_{j=1}^{Q} x_{ij}^{k} = a_i, \ i = 1 \ldots Z \\ \sum_{i=1}^{Z} x_{ij}^{k} = b_j, \ j = 1 \ldots Q \\ \sum_{i=1}^{Z} a_i = \sum_{j=1}^{Q} b_j \\ a_i \geq 0, \ i = 1 \ldots Z; \ b_j \geq 0, \ j = 1 \ldots Q \end{cases}, \quad (1.8)$$

where:

x_{ij}^{k} the number of M&R services for the *j*-th vehicle of the *k*-th transport company ($x_{ij}^{k} \geq 0$) that are assigned to the *i*-th MSC or the MTC repair unit

Three alternative solutions of the problem can be found, where values of the $\mathbf{X_{ZQ}}$ matrix minimize the following linear objective functions:

$$F_{KDC} = \sum_{i=1}^{Z} \sum_{j=1}^{Q} KDC_{ij} \cdot x_{ij}^{k} \quad \rightarrow \quad \min, \quad (1.9)$$

$$F_{KD} = \sum_{i=1}^{Z} \sum_{j=1}^{Q} KD_{ij} \cdot x_{ij}^{k} \quad \rightarrow \quad \min, \quad (1.10)$$

$$F_{C} = \sum_{i=1}^{Z} \sum_{j=1}^{Q} C_{ij} \cdot x_{ij}^{k} \quad \rightarrow \quad \min, \quad (1.11)$$

In the presented formulation, the M&R distribution problem can be solved by using the Hungarian algorithm.

After obtaining three alternative distribution plans, the admissibility of the solutions should be checked by the servicing costs and the coefficient of downtime values:

$$C_{PL}^{(v)} = \sum_{i=1}^{Z} \sum_{j=1}^{Q} C_{ij} \cdot x_{ij}^{k} \leq C_A, \quad (1.12)$$

$$KD_{PL}^{(v)} = \sum_{i=1}^{Z} \sum_{j=1}^{Q} KD_{ij} \cdot x_{ij}^{k} \leq KD_A, \quad (1.13)$$

where:

$C_{PL}^{(v)}$ total M&R costs of the transport company's vehicles according to the v-th distribution plan,

C_A allowable costs for M&R of vehicles,

$KD_{PL}^{(v)}$ the resulting downtime of the MTC vehicles according to the v-th distribution plan,

KD_A the admissible downtime coefficient

To choose the most appropriate distribution plan, we propose to use the following criterion Δ_v:

$$\Delta_v = w_c \cdot \left(C_A - C_{PL}^{(v)}\right) + w_d \cdot \left(KD_A - KD_{PL}^{(v)}\right) \rightarrow \max, \tag{1.14}$$

where:

w_c, w_d the weight coefficients reflecting the significance of the M&R costs and the downtime coefficient for a decision-maker

If a valid plan was not obtained, the initial data should be changed and the parameters of MSCs in the region reevaluated.

If there are more than one admissible plan, the most appropriate for the distribution conditions is the plan with the maximum value of the criterion Δ_v.

At the final stage, based on the results of the chosen distribution plan, recommendations on the assignment of M&R services for the MSC and MTC repair units are being developed.

1.4 Case Study and Discussion

The experimental studies were conducted by using the case of the transport and service companies located in Dergachi and Vesele settlements (Kharkiv region, Ukraine). The data on the forecasted M&R services in October 2019 were used.

The available equipment for the maintenance of service companies located in the chosen settlements was identified. The preliminary selection of MSCs was carried out in accordance with the brand composition of transport companies' vehicles, with their position at the M&R market, and with the ability to provide warranty obligations for provided services. The production capacities of MSCs were calculated based on the obtained information (see Table 1.1).

Based on the data on the service lifecycle for the type of vehicles used at transport companies in the regions, the annual demand for M&R services was estimated as the sum of maintenances and repairs for each vehicle. For each vehicle model used by transport companies, the costs of services at each of the considered MSCs were calculated, and the values of downtime coefficients were assessed.

Table 1.1 Production abilities of the servicing companies

Servicing company	Estimated production capacity [veh./year]	Location
Veles	129	Dergachi
UAZ Service	158	Dergachi
Sokirko	135	Dergachi
AutoService 67	144	Dergachi
Dergachi service	193	Dergachi
MIX Service	114	Vesele
Pit-Stop	93	Vesele
Garage service	121	Vesele

The obtained matrices were used as objective function coefficients for solving the resource allocation problem for each of the objectives presented in (1.9)–(1.11).

Optimal plans for the distribution of services for transport companies in the region were obtained by applying the Hungarian algorithm to the prepared data. The values of the objective functions assessed for each of the plans are shown in Table 1.2.

From the alternative distributions, plan 3 of the M&R services was chosen based on the criterion (1.13). The obtained results of the distribution of services for Dergachi and Vesele settlements are presented in Tables 1.3 and 1.4.

Table 1.2 Characteristics of the obtained distribution plans

Distribution plan v	Value of the objective function		
	F_{KDC}	F_{KD}	F_C
Plan 1 ($F_{KD} \to \min$)	10799.43	32.404	444.22
Plan 2 ($F_{KDC} \to \min$)	10798.45	33.012	445.58
Plan 3 ($F_C \to \min$)	10798.98	33.238	443.99

Table 1.3 Distribution plan of M&R services for the Dergachi settlement

Servicing company	Vehicle models				
	UAZ 3151, 3303	VW Amarok	VAZ 2123	GAZ 2705, 3307	Bogdan A092
Veles	26	–	–	–	–
UAZ service	–	–	10	–	–
Sokirko	–	–	–	–	16
AutoService 67	–	–	–	8	–
Dergachi service	–	15	–	–	–

Table 1.4 Distribution plan of M&R services for the Vesele settlement

Servicing company	Vehicle models			
	UAZ 3151, 3303	VAZ 2123	GAZ 2705, 3307	Bogdan A092
MIX service	–	–	–	11
Pit-Stop	47	11	–	–
Garage service	–	–	17	–

The consideration in the proposed method of time for moving of vehicles the to the places of service, of a load of servicing companies by its own customers, and of the companies' ability to service vehicles allowed us to reduce the vehicles' downtime due to maintenance by 10%. In turn, the coefficient of technical readiness of the transport companies' fleet has increased by about 12%.

1.5 Conclusions

Comparing to the existing methodologies, the proposed method considers the possibility of receiving group M&R requests by motor service companies. It assesses the ability of motor service companies to service vehicles depending on the flow of existing customers. Assessment of motor service companies considers not only the duration of servicing operations but also the distance to the vehicles to be serviced.

The presented method allows considering transport companies in the region as a unified system for which the most expedient variant of involvement of motor service companies can be forecasted. The proposed method, depending on the capabilities of the region, allows choosing a strategy to maintain the technical readiness of the transport companies' fleet by developing their own repair units or by involving third-party motor service companies.

It should be noted that the proposed method of ensuring the technical readiness of companies' vehicles due to the region's capabilities demands preliminary preparation of a considerably big amount of initial data. However, the optimized distribution plan based on this data can be obtained without time-consuming calculations as it represents the solution of the assignment problem in its classical formulation.

References

1. Martorell S, Villamizar M, Carlos S, Sánchez A (2010) Maintenance modeling and optimization integrating human and material resources. Reliab Eng Syst Safety 95(12):1293–1299
2. Yang J-H, Guo L (2020) The optimization model of comprehensive maintenance plan based on multi demand service supply chain. Adv Intell Syst Comput 1117:1783–1802
3. Deblaere F, Demeulemeester E, Herroelen W, van De Vonder S (2007) Robust resource allocation decisions in resource-constrained projects. Decis Sci 38(1):5–37

4. Kim S, Frangopol DM (2017) Efficient multi-objective optimisation of probabilistic service life management. Struct Infrastruct Eng 13(1):147–159
5. Allah Bukhsh Z, Stipanovic I, Doree AG (2020) Multi-year maintenance planning framework using multi-attribute utility theory and genetic algorithms. Eur Trans Res Rev 12(1), article 3
6. Aghezzaf E-H, Sitompul C, Najid NM (2010) Models for robust tactical planning in multi-stage production systems with uncertain demands. Comput Oper Res 37(5):880–889
7. Garcia-Herreros P, Zhang L, Misra P, Arslan E, Mehta S, Grossmann IE (2016) Mixed-integer bilevel optimization for capacity planning with rational markets. Comput Chem Eng 86:33–47
8. Ren J, Chen C, Zhang X (2018) Multi-scenario optimisation model for reusable logistics resource allocation. J Southwest Jiaotong Univ 53:1270–1277
9. Alizadeh Afrouzy Z, Nasseri SH, Mahdavi I, Paydar MM (2016) A fuzzy stochastic multi-objective optimization model to configure a supply chain considering new product development. Appl Math Model 40(17–18):7545–7570
10. Medina-González S, Pozo C, Corsano G, Guillén-Gósalbez G, Espuña A (2017) Using Pareto filters to support risk management in optimization under uncertainty: Application to the strategic planning of chemical supply chains. Comput Chem Eng 98:236–255
11. Naumov V, Starczewski J (2019) Choosing the localisation of loading points for the cargo bicycles system in the Krakow Old Town. Lecture notes in networks and systems, vol 68, pp 353–362
12. Naumov V, Taran I, Litvinova Ya, Bauer M (2020) Approach to optimization of multimodal transport hub resources for material flow service. Sustainability 12(16), article 6545
13. Tubis A, Werbińska-Wojciechowska S (2018) The scope of the collected data for a holistic risk assessment performance in the road freight transport companies. In: Zamojski W, Mazurkiewicz J, Sugier J, Walkowiak T, Kacprzyk J (eds) Advances in dependability engineering of complex systems. DepCoS-RELCOMEX 2017. AISC, vol 582. Springer, Cham, pp 450–463
14. Sawicka H, Bodziony P, Sawicki P (2017) Selection of technological vehicles to the geological and mining conditions with an application of stochastic group decision aiding method. In: Proceedings of the carpathian logistics congress, pp 178–184
15. Sierpiński G, Macioszek E (2020) Equalising the Levels of Electromobility Implementation in Cities. In: Mikulski J (ed) Research and the future of telematics, vol 1289. Communications in computer and information science. Springer, Heidelberg, pp 165–176
16. Macioszek E, Sierpiński G (2020) Charging stations for electric vehicles-current situation in Poland. In: Mikulski J (ed) Research and the future of telematics, vol 1289. Communications in computer and information science. Springer, Heidelberg, pp 124–137
17. Albey E, Bilge U, Uzsoy R (2017) Multi-dimensional clearing functions for aggregate capacity modeling in multi-stage production systems. Int J Prod Res 55(14):4164–4179

Chapter 2
Analysis of the Air Quality in Considering the Impact of the Atmospheric Emission from the Urban Road Traffic

Iryna Lynnyk, Kateryna Vakulenko, and Elena Lezhneva

2.1 Introduction

The risk of environmental pollution is assessed by the level of its possible negative impact on the atmosphere, soils, groundwater and surface water, vegetation, animals and humans [1–5]. Atmospheric air pollution comes from mobile and stationary sources of pollution. Stationary sources include emissions from large industrial enterprises, machine-building, coke and chemical industries. One of the main sources of pollution is thermal power plants.

Mobile sources include emissions from vehicles. The main source of emissions and emissions on highways are vehicles. The impact of traffic flow on the environment is considered as the sum of the impacts of single cars. The ecological danger of a single car is determined not only by its design, but also by the mode of movement.

The risk of chemical pollution due to the operation of the road complex is assessed by the level of its possible negative impact on the atmosphere and people. Therefore, an urgent problem is the assessment of air quality.

I. Lynnyk · K. Vakulenko (✉)
O.M. Beketov National University of Urban Economy in Kharkiv, Kharkiv, Ukraine
e-mail: vakulenko.e@ukr.net

E. Lezhneva
Kharkiv National Automobile and Highway University, Kharkiv, Ukraine

E. Macioszek and G. Sierpiński (eds.), *Research Methods in Modern Urban Transportation Systems and Networks*, Lecture Notes in Networks and Systems 207, https://doi.org/10.1007/978-3-030-71708-7_2

2.2 Analysis of the Air Quality on the Example of the Kharkiv City of Ukraine

In Kharkiv, according to the Main Department of Statistics in the Kharkiv region, emissions of air pollutants from stationary sources in 2019 amount to 4,013 thousand tons [6] (Fig. 2.1). The main air pollutants of the city are Thermal Power Station (TPS)-3, Kharkiv Tractor Plant, State Plant "Plant named after Malyshev", Closed Joint Stock Company "Kharkiv Coke Plant" [7].

Atmospheric pollution by vehicle emissions ranks second after energy production sector due to the constant increase in the number of vehicles.

The main causes of intense air pollution from mobile sources are:

- annual increase in the total number of vehicles,
- operation of technically obsolete car fleet,
- low quality of fuels and lubricants,
- unsatisfactory condition of highways,
- lack of detour routes,

Fig. 2.1 The level of air pollution in the city of Kharkiv as of April 2020 (the circles show the air quality indices) *Source* Own research based on [7]

- insufficient capacity of the road transport network, which was formed in the conditions of the existing buildings, especially in the central part of the city,
- poor organization of traffic,
- lack of road junctions, underground pedestrian crossings, etc.,
- unsatisfactory condition of the road surface of the roadway.

Exhaust gases from cars emit about 280 different harmful substances, among which carcinogenic benzo[a]pyrene, oxides of nitrogen, lead, mercury, oxides of carbon and sulfur, soot, hydrocarbons, aldehydes, gasoline are especially dangerous., vanadium, kerosene, cobalt, copper, nickel, lead, etc. The presence of harmful substances in the exhaust gases is due to the variety of fuels, additives and oils, fuel combustion conditions, engine operation, its technical condition, driving conditions, etc.

The presence of harmful substances in the exhaust gases leads to pollution of roads and adjacent to the road surface of land and forest belts, and in case of precipitation of surface and groundwater, harms human health, and in some cases leads to serious diseases. The data presented in [8] indicate that the specific emissions of pollutants and greenhouse gases from the consumption of one ton of fuel for gasoline engines contain the largest number of harmful compounds. One solution is to gradually replace the fleet of conventionally powered vehicles with electric vehicles [9–12].

Normal for the human body is the complete absence of air pollutants. However, modern technologies cannot ensure the purity of the atmosphere due to existing methods of energy conversion. Therefore, when deciding on the protection of the atmosphere, the agreement on its permissible pollution is taken into account, which is estimated by the value of the permissible concentration of pollutants in the atmosphere.

According to the recommendations of the World Health Organization (WHO), depending on the concentration of CO (mg/m3), the following degrees of air pollution are distinguished: light—0–7; weak—8–13; moderate—14–27; significant—28–40; serious—41–53; very serious—54–67; threatening—68–80; dangerous—more than 80.

Mild air pollution has no direct or indirect negative impact on humans.

Atmospheric pollution is legally limited by sanitary regulations protection of atmospheric air by the introduction of maximum permissible concentrations of harmful substances. These concentrations are different for different conditions of human activities. Thus, the concentration of CO is permissible in populated areas taken equal to 3 mg/m3, which falls within the range of mild pollution according to WHO recommendations. In the range of moderate pollution is the maximum permissible concentration (MPC) which corresponds to the maximum single concentration of CO in the air of the working area (Table 2.1).

Comparison the accepted maximum of allowable concentrations of chemicals in Ukraine, the United States and the European Union, conclusion is the Ukrainian standards are stricter (Table 2.2) [8, 13, 14].

Atmospheric air quality is assessed by the quality index (AQI). The air pollution index is a simplified indicator and is usually calculated for the five most significant

Table 2.1 Content of some pollutants in the air of Kharkiv. *Source* [6]

Name of polluting substances	Average annual content [mg/m^3]	Average daily MPC [mg/m^3]	Maximum one-time MPC [mg m^3]	Maximum content [mg/m^3]
Road dust	0.090	0.150	0.500	1.600
Sulphur dioxide (SO_2)	0.007	0.050	0.500	0.029
Sulfates	0.000	–	–	0.010
Carbon monoxide (CO)	2.800	3.000	5.000	12.00
Nitrogen dioxide (NO_2)	0.020	0.040	0.200	0.120
Nitrogen monoxide (NO)	0.020	0.060	0.400	0.050
Hydrogen sulfide	0.001	–	0.008	0.003
Phenol	0.001	0.003	0.010	0.006
Soot	0.030	0.050	0.150	0.300
Ammonia	0.000	0.040	0.200	0.040
Formaldehyde	0.002	0.003	0.035	0.017

Table 2.2 Comparison of some maximum permissible concentrations of chemicals in Ukraine, the USA and the European Union

Substance	MPC [mg/m^3]		
	Ukraine	USA	European Union
SO_2	0.050	0.075	0.125
NO_2	0.085	0.053	0.200
Pb	0.001	0.00015	0.0005
CO	5.000	10	10

concentrations of substances that determine the total air pollution: benzo[a]pyrene, formaldehyde, phenol, ammonia, nitrogen dioxide, carbon disulfide. In Fig. 2.1. the state of the atmospheric air of the city of Kharkiv is presented according to the observations of the Kharkiv Regional Center for Hydrometeorology [7].

Analyzing the level of air pollution in the city, conclusion is that it is extremely high.

Average air quality indices in the city of Kharkiv for the period from August 3 to September 2, 2020 are shown in Fig. 2.2 [7].

The graph (Fig. 2.2) shows that air pollution has decreased due to the introduction of quarantine measures, non-operating enterprises and a decrease in the number of vehicles on the city streets.

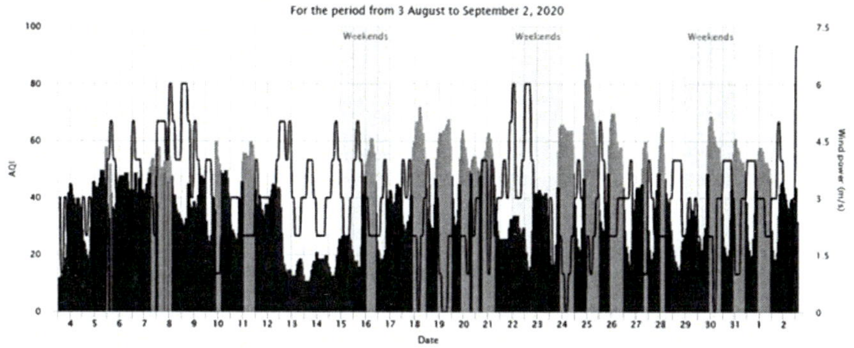

Fig. 2.2 The average air quality index in Kharkiv from April 9 to May 9, 2020. *Source* Own research based on [7]

2.3 Assessment of Air Pollution of the System "Road-Environment"

Road transport is one of the main sources of environmental pollution. Emissions from motor vehicles are the main cause of declining air quality in cities in Europe, China, Asia and other countries. In recent years, some progress has been made in developing methods and measures to reduce emissions of air pollutants [1–8, 13, 14]: methods for assessing air pollution and the effectiveness of improving air quality in the implementation of measures for traffic management and organization in cities, parking management, promotion public transport use by the citizens, bicycles, pedestrians, measures to minimize travel time for work and leisure, measures for land use and development (organization of low-emission zones, i.e. areas within the city or region where vehicles must meet certain emission standards or are required to pay a fee), integrated urban freight transport, measures to improve transport and fuel technologies, measures on tax schemes and fees, etc. [13, 15].

In [16] it is proposed to reduce the level of CO2 through the introduction of "environmental driving" by optimizing road infrastructure. The principle of "environmental driving" is aimed at detecting incorrect positions of Misplaced Speed-sectioning Positions (MSP) relative to the Starting Point of Deceleration (SPD), under which the speed sectioning is a sequence of changes in speed along a certain route. The main purpose of "environmentally friendly driving" is to minimize the movement of the car in the start-stop mode and ensure the continuity of traffic flow, while the dynamics of fuel consumption is reduced, which leads to lower emissions. Quantitative assessment of fuel consumption due to incorrectly installed road signs for traffic flow, taking into account its composition, allows to model the driver's behavior and fuel consumption under different conditions and reduce fuel consumption.

The formation of Environmental Areas within the city to improve air quality is given attention in [17]. In this chapter, special attention is paid to the implementation and management of the street trees and the selection of trees, with appropriate leaf characteristics, to improve the environment. At the same time, the organization of such ecological zones will promote the development of pedestrian and bicycle traffic, by restricting traffic for cars.

The social impact of pollution levels from mobile sources is a key element in urban planning and traffic planning, so paper [18] proposed an assessment of the social and medical consequences of air pollution in Madrid (Spain). The proposed methodology for estimating emissions in cash, taking into account the costs associated with loss of working time and mortality, with reference to CORINAIR data [14] on information on emissions into the atmosphere by different modes of transport related to acid rain [14].

To assess air pollution, air monitoring is used, which depends on direct measurements, biological measurements, where biological markers are used to assess the impact [19–21], biotesting methods, biodiagnostics. Today, when the contradiction between the economy and the environment is exacerbated, it is important that assessment methods can not only give an objective picture of the state of the atmosphere, but also be available in material terms.

Computer modeling [22] such as COPERT and others has become widely used in the assessment of air pollution. This technique is more complicated and accurate, but collecting the necessary input requires a lot of resources and time [23]. It should be noted that some models and databases calculate only emissions from one source under different conditions, others estimate emissions from one transport node, or simulate total emissions for a specific area. Modeling the amount of pollutant emissions on the roads can be both static and dynamic, depending on the application and calculation method [22].

In [18], a mathematical model for estimating emissions on the street-road network NEMO is proposed, which includes the following components: spatial components/transport infrastructure (data on the geometry of the selected area, land use, road network); transport component (intensity and composition of traffic flow, type of car, type of engine, etc.); component of the level of pollution, taking into account weather conditions. This model was tested in Wroclaw, Poland. The results of the analysis confirm that large congestion within the study area leads directly to the formation of smog due to significant emissions of NOx, CO and particulate matter. The use of this model requires detailed preparation and time, so in its calculation, assumptions and simplifications were made, which could affect the accuracy of modeling.

Along with the main share of harmful emissions into the atmosphere, the development of the road complex and, as a consequence, the increase in traffic intensity of the highway led to a real danger of environmental quality as a result of abnormal changes in sound characteristics (frequency, volume) in populated areas and other places. With the increase in the number of vehicles and the speed of their movement on the streets of industrial cities, the world community has identified noise as one of the main factors that worsen the living standards of people in cities.

It is impossible to physically avoid noise pollution; it is only possible not to notice it subjectively. Emotional and physical stress associated with constant noise discomfort leads to noise stress. Therefore, the problem of noise pollution by highways is no less relevant than chemical, as studies identify new aspects of the negative acoustic impact on the health of residents of large cities [24].

Modern methods are based on technical and economic criteria and do not provide for systematic optimization of the volume of work for each of the measures.

The analysis of existing methods has shown that they are unsystematic. In addition, attempts to take into account the negative impact of the road on biocenoses, landscapes within the technical and economic approach are limited by the possibilities of long-term forecasting methods. As a result, current rather than future technological and economic opportunities of society become the basis on which evidence of the effectiveness of environmental measures is built.

In contrast to the previously proposed methods for solving the problem of assessing air pollution, this chapter focuses on the systematic optimization of measures taking into account the technical requirements of cars, drivers and sanitary requirements of pedestrians. Therefore, it is necessary to find for the system "road–environment" the optimal tactics of behavior—a rule of change that maximizes the existence of the system.

Emissions of air pollutants are calculated by the formula:

$$M_i = 0,0548 \cdot M_x \cdot \rho_n \cdot X_i \cdot Q_i \cdot \tau, \tag{2.1}$$

where: M_i—emissions of harmful substances, g/km; M_x—molecular weight of toxic substances, g/mol; X_i—content of harmful substances [%]; Q_i—fuel consumption, t/year, or kg/h, or g/s; τ—excess air ratio.

Fuel consumption and the amount of harmful emissions from cars can be calculated by the formulas of Govorushchenko [25]:

$$Q_1 = \frac{1}{\eta_{mp}}\left[A \cdot i_k + B \cdot i_k^2 + C\begin{pmatrix} C_{1a} \cdot \psi + 0,077 \cdot k \cdot F_k \cdot V^2 + \\ + 0,1 \cdot C_{1a} \cdot \dot{V} \end{pmatrix}\right], \tag{2.2}$$

where: ψ—coefficient of road resistance; k—air resistance coefficient, Hc2м-4; F_k—frontal area of the car, m2; V—car speed, m/s; $\dot{V}$—acceleration, m/s2; δ—the coefficient of accounting for rotational masses.

$$A = \frac{7,95 \cdot a \cdot V_n \cdot i_o}{H_H \cdot \rho_n \cdot r_k}; \quad B = \frac{0,69 \cdot b \cdot V_n \cdot S_n \cdot i_o}{H_H \cdot \rho_n \cdot r_k^2}; \quad C = \frac{100}{H_H \cdot \rho_n \cdot \eta_{mp}};$$

where: V_n—engine cylinder capacity, H_H—lower heat of combustion of fuel, kJ/kg ($H_H = 44000$ kJ/kg—for gasoline, $H_H = 43000$ kJ/kg—for diesel fuel); ρ_n—fuel density, g/cm3 ($\rho_n = 0.740$—for gasoline, $\rho_n = 0.825$—for diesel fuel); S_n—piston stroke, м; i_k—gear ratio; i0—gear ratio of the main transmission; η_{mp}—transmission efficiency ($\eta_{mp} = 0.875$—for a car with one drive axle, $\eta_{mp} = 0.825$—for a car with two leading axles); a, b—constant coefficients ($a = 48$ kPa—for diesels, $b = 16$ kPa—for diesels, $a = 45$ kPa sm-1—for injector engines, $b = 13$ kPa sm-1—for

Table 2.3 Indicators of road surface equality

Coating	Condition		
	Excellent	Good	Unsatisfactory
Asphalt and cement concrete	50–75	150	300
Crushed stone and gravel	200	350–400	800–900
Pavement (stone)	300	500	1000

injector engines, r_k is equal to from 0.03 to 0,05—for cars, r_k is equal to from 0.05 to 0.07—for trucks.

The coefficient of road resistance ψ is estimated (according to O.K. Birul) [26]:

$$\psi = 0,01 + \lambda_n \cdot S_n \cdot V^2 \cdot 10^{-8} + i, \tag{2.3}$$

where: λ_n—coefficient depending on the design of the chassis of the car; $\lambda_n = 4.0$ for cars,$\lambda_n = 5.5$ for trucks; S_n—indicator of equality of a road covering (Table 2.3); V—current car speed, km/h; i—longitudinal slope, ‰.

Substituting numerical values M_{x} in formula (2.1) leads to get the following dependences to determine the emissions of various substances into the atmosphere.

For CO:

$$M_{CO} = 1,53 \cdot \rho_n \cdot X_{CO} \cdot Q_1 \cdot \tau. \tag{2.4}$$

Where

$$X_{CO} = 61,3 - 144 \cdot \tau + 53 \cdot \tau^2. \tag{2.5}$$

For NO:

$$M_{NO} = 1,64 \cdot \rho_n \cdot X_{NO} \cdot Q_1 \cdot \tau. \tag{2.6}$$

Where

$$X_{NO} = -3,67 + 7,88 \cdot \tau - 3,88 \cdot \tau^2. \tag{2.7}$$

For NO_2:

$$M_{NO_2} = 2,52 \cdot \rho_n \cdot X_{NO_2} \cdot Q_1 \cdot \tau, \tag{2.8}$$

where

$$X_{NO_2} = -3,67 + 7,88 \cdot \tau - 3,88 \cdot \tau^2. \tag{2.9}$$

For C_6H_{14}:

$$M_{CH} = 4,7 \cdot \rho_n \cdot X_{CH} \cdot Q_1 \cdot \tau, \tag{2.10}$$

where

$$X_{CH} = 0,922 - 1,677 \cdot \tau + 0,77 \cdot \tau^2. \tag{2.11}$$

The concentration of pollutants in the air C_i, in milligrams per cubic meter, at a distance «x» from the source of pollution is calculated by the Bozanke-Pearson formula. The solution of this equation in the case when the traffic flow is considered as a linear source, is presented as follows [27]:

$$C_i = \frac{\Delta M_i \cdot 1000 \cdot \eta}{W_B \cdot p \cdot x} \exp\left[-\frac{H}{p \cdot x}\right] + C_f, \tag{2.12}$$

where x—the distance from the source of pollution to the building or reserve-technological zone of the road, m; M_i—emission of a pollutant, g/s on the running length of the linear source 1 m; W_B—wind speed perpendicular to the direction of the road, m/s; H—the height of the source above the carriageway (0.4 m—for passenger traffic; 0.5 m—for mixed traffic flow; 0.6 m—for freight traffic flow); p–coefficient taking into account the influence of the scattering angle of the pollutant in the vertical plane due to the turbulence of the atmosphere (p take from 0,05 to 0,30); Δ—coefficient of the implementation and management of street trees impact; η—building impact factor; C_f—background concentration of the pollutant in the air, mg/m 3.

Building impact factor η is determined:

$$\eta = 1 + 0.044 \cdot (x - B_z + L_T) + 0.0013 \cdot (x - B_z + L_T)^2, \tag{2.13}$$

where: B_z—distance from the source of pollution to the building, m; L_T—length of aerodynamic shadow, m:

$$L_T = H_z \text{ under } H_z \leq L_{ш} \tag{2.14}$$

$$L_T = L_{ш} \text{ under } H_z \geq L_{ш} \tag{2.15}$$

where: $H_з$—building height, m; $L_{Ш}$—building width, m.

The proposed method for determining emissions and concentrations of pollutants in contrast to existing ones is based on the system optimization. It can be used to prescribe appropriate measures at different stages of the system "road–environment" operation.

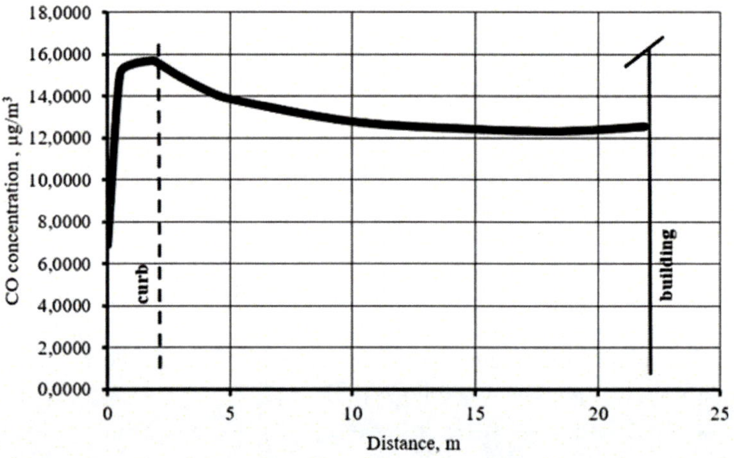

Fig. 2.3 CO scattering

The analysis of the ecological situation on the experimental section of Jubilee Avenue in the city of Kharkiv was carried out on the basis of air pollution by harmful emissions from vehicles.

The calculation of concentrations of air pollutants by harmful emissions from vehicles was performed using the model for estimating air pollution (9) and the program "ECO NORM", developed according to research by Lynnyk [28]. The traffic intensity on Jubilee Avenue was 3.5 thousand cars per hour, the wind speed was 1.1 m/s. The composition of the traffic flow: cars—78.07%; minibuses—3.6%; light trucks—4.95%; medium trucks—1.89%; heavy trucks—5.6%; buses—6.84%. The average speed of the traffic flow was 30.0 km/h. Indices of equality of the carriageway on the experimental site did not exceed 50 cm/km. The longitudinal slope of the road is 12‰. The results of the calculations are presented at Figs. 2.3, 2.4, 2.5, 2.6, 2.7 and 2.8.

The graphs show that the concentrations in the working area: carbon monoxide (CO) exceed the MPC by 3.85 times; nitric oxide (NO) exceeds the MPC by 3 times; hexane (CH) exceeds the MPC by 4 times; lead (Pb) exceeds the MPC by 5 times; sulfur dioxide (SO2) exceeds the MPC by 3 times, soot exceeds the MPC by 6 times.

Concentrations of pollutants at the building level: carbon monoxide (CO) exceed the MPC by 2.5 times; nitric oxide (NO) exceeds the MPC by 1.5 times; hexane (CH) exceeds the MPC by 2 times; lead (Pb) exceeds the MPC by 5 times; sulfur dioxide (SO2) exceeds the MPC by almost 2.5 times, soot exceeds the MPC by 5 times.

Verification of the adequacy of the model for estimating air pollution (9) was performed based on the average approximation error by comparing the calculated concentrations of air pollutants with the actual (Fig. 2.9). The average approximation error indicator is 1.168%, which satisfies the conditions of adequacy and indicates the coincidence of theoretical and actual values. The air quality at Jubilee Avenue

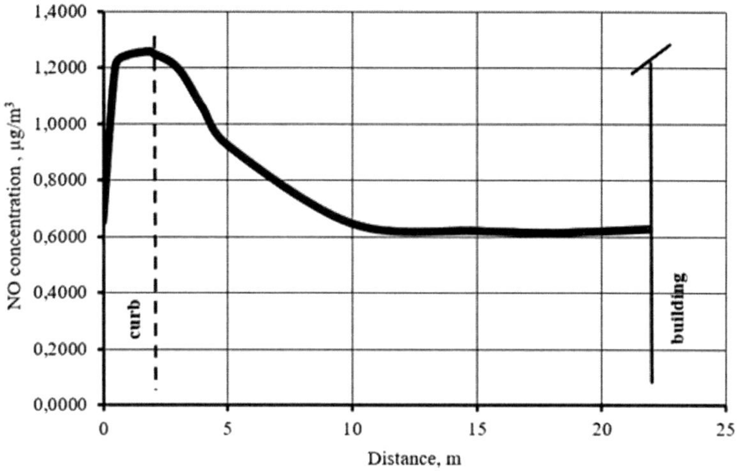

Fig. 2.4 NO scattering

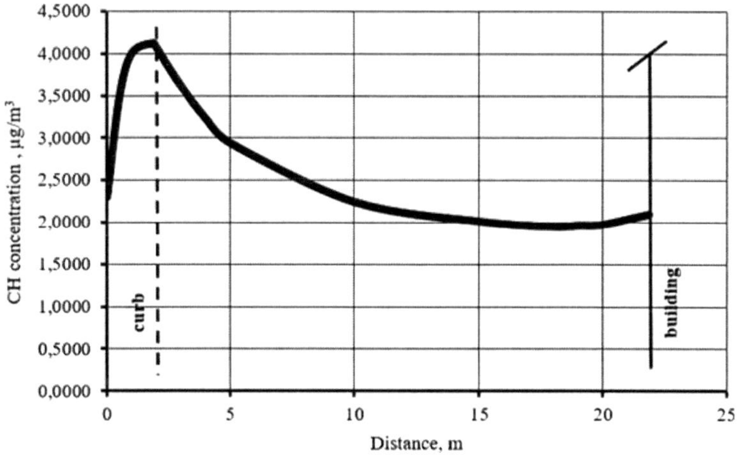

Fig. 2.5 CH scattering

is of low quality and requires measures to improve it regarding to the results of the method implementation.

To reduce gassiness in the areas adjacent to Jubilee Avenue, the following urban planning measures are proposed: to limit the movement of heavy goods vehicles in this area; traffic speed regulation; use of elements of traffic organization; maximum implementation and management of street trees at the areas adjacent to the street; use of noise-protective and gas-protective species of trees and shrubs in landscaping; replacement of existing window fillings with plastic double-glazed windows, which are protective against harmful factors. Also, a perspective area of protection of the

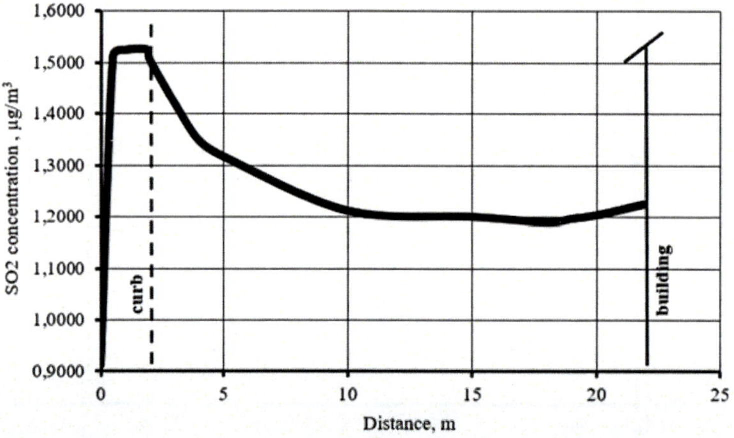

Fig. 2.6 SO_2 scattering

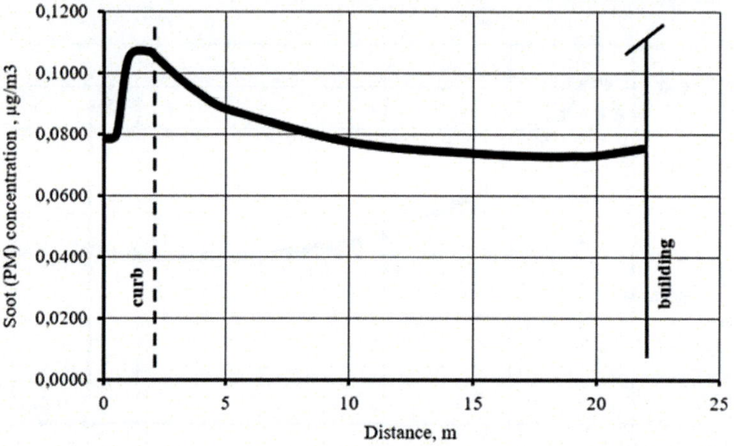

Fig. 2.7 Soot scattering

residential areas near the highways is the use of acoustic screens [24]. The advantages of using acoustic screens in comparison, for example, with greenery, should be noted as the constant efficiency, regardless of the time of year, leaf density. In addition, the effectiveness of acoustic screens comes from the moment of their installation, while to achieve a certain noise protection efficiency of greenery takes a long time until the trees and shrubs reach a certain height and other characteristics.

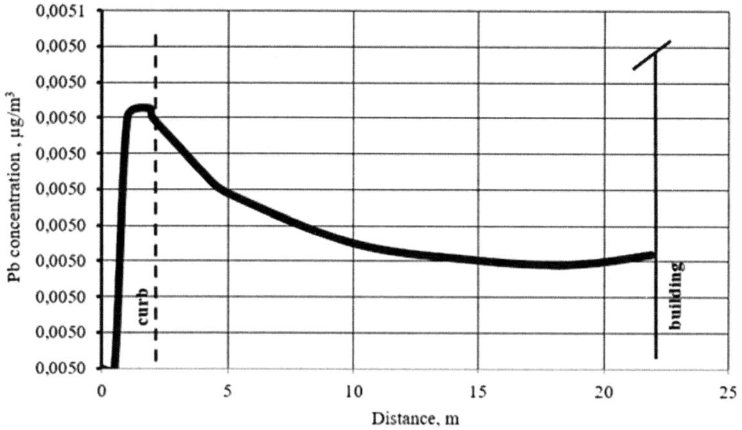

Fig. 2.8 Lead scattering

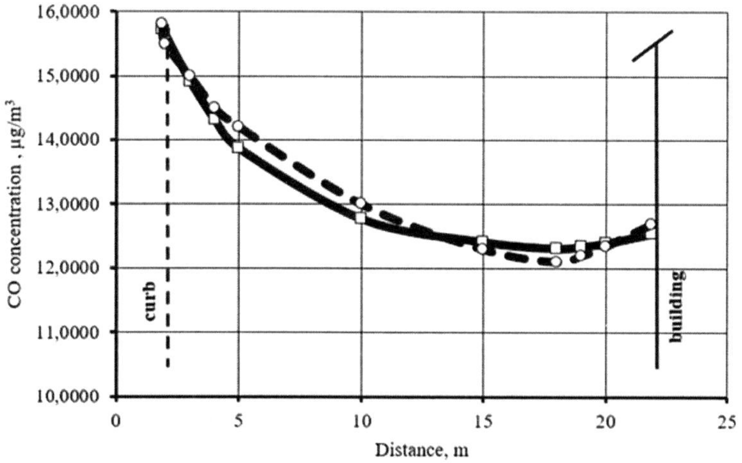

where:

- - - - - actual values; ——— - theoretical values.

Fig. 2.9 Theoretical and actual values of CO scattering at Jubilee Avenue

2.4 Conclusions

The conducted research of influence the atmospheric emission from the urban road traffic on air quality in the city presented in the chapter there was formulate the following conclusions:

- as a result of the analysis of the current state of atmospheric air in Kharkiv and Kharkiv region, it was found that atmospheric air pollution comes from mobile and stationary sources of pollution. One of the main sources of pollution is thermal power plants,
- the most polluted territories of the city of Kharkiv were identified, especially the central district and the area of the railway station, and the causes of pollution were established,
- as a result of comparison of the accepted maximum admissible concentrations of chemical substances in Ukraine, the USA and the countries of the European Union it has been revealed that the Ukrainian norms are stricter,
- the proposed method of estimating air pollution in contrast to the existing ones is based on the system optimization. It can be used to take appropriate measures at different stages of the system "road–environment",
- as a result of calculations of concentrations of pollutants from motor vehicles at Jubilee Avenue, it has been found out that for almost all the substances their concentrations exceed the MPC by more than 3 times,
- the main measures to improve the state of atmospheric air at Jubilee Avenue in Kharkiv have been proposed.

References

1. Ukraine, Derzhavni budivel′ni normy D.B.N A. 2.2-1-2003. https://zakon.rada.gov.ua/rada
2. Office of Air Quality Planning and Standards, U.S. Environmental Protection Agency: Emission assessment report. https://www3.epa.gov
3. Mehraj SS, Bhat GA (2013) Cement factories, air pollution and consequences. University of Kashmir, India
4. Abimbola AF, Kehinde-Phillips OO, Olatunji AS (2007) The Sagamu cement factory, SW Nigeria: Is the dust generated a potential health hazard? SEGH 29(2):163–167
5. Ball DJ, Hamilton RS, Harrison RM (1991) The influence of highway-related pollutants on environmental quality. Stud Environ Sci 44:1–47
6. Holovne upravlinnia statystyky u Kharkivskii oblasti. http://kh.ukrstat.gov.ua
7. SaveEcoBot Air pollution level in the city Kharkiv. https://www.saveecobot.com
8. United States Environmental Protection Agency. National Ambient Air Quality Standards (NAAQS) for PM. https://www.epa.gov
9. Sendek-Matysiak E (2019) The role and importance of electric cars in shaping a sustainable road transportation system. In: Siergiejczyk M, Krzykowska K (eds) Research methods and solutions to current transport problems, AISC, vol 1032. Springer, Cham, pp 381–390
10. Alrawi F (2017) The importance of intelligent transport systems in the preservation of the environment and reduction of harmful gases. Transp Res Procedia 24:197–203
11. Sierpiński G, Macioszek E (2020) Equalising the levels of electromobility implementation in cities. In: Mikulski J (ed) Research and the future of telematics, vol 1289. Communications in computer and information science. Springer, Heidelberg, pp 165–176
12. Macioszek E, Sierpiński G (2020) Charging stations for electric vehicles-current situation in Poland. In: Mikulski J (ed) Research and the future of telematics, vol 1289. Communications in computer and information science. Springer, Heidelberg, pp 124–137
13. Air Quality and Urban traffic in the EU: best practices and possible solutions. https://www.europarl.europa.eu

14. EMEP/EEA air pollutant emission inventory guidebook 2019. https://www.eea.europa.eu
15. Elsom DM (1997) Effectiveness of traffic management measures in improving air quality in European cities. Trans Ecol Environ 15:59–68
16. Huang X, Tang G, Zhang J (2021) Characteristics of PM2.5 pollution in Beijing after the improvement of air quality. J Environ Sci 100:1–10
17. Ortolani C, Vitale M (2016) The importance of local scale for assessing, monitoring and predicting of air quality in urban areas. Sustain Cities Soc 26:150–160
18. Monzón A, Guerrero MJ (2004) Valuation of social and health effects of transport-related air pollution in Madrid (Spain). Sci Total Environ 334:427–434
19. Hoek G (2017) Methods for assessing long-term exposures to outdoor air pollutants. Current Environ Health Rep 4(4):450–462
20. Watson AY, Bates RR, Kennedy D (1988) Assessment of human exposure to air pollution: methods, measurements, and models. In: Air pollution, the automobile, and public health. National Academies Press, US
21. Watson AY, Bates RR, Kennedy D (2012) Air pollution, the automobile, and public health. Sponsored by The Health Effects Institute
22. Skrętowicz M, Galas-Szpak A (2018) Analysis of the impact of motor vehicles on the air quality on the example of Legionow Square in Wroclaw. In: International automotive conference (KONMOT2018). IOP Publishing, pp 1–10
23. Bellasio R, Bianconi R, Corda G, Cucca P (2007) Emission inventory for the road transport sector in Sardinia (Italy). Atmos Environ 41(4):677–691
24. Lezhneva E, Vakulenko K (2019) Assessing of traffic noise pollution of road transport in urban residential. Roman J Trans Infrastruct 8(1):34–52
25. Hovorushchenko N (1990) Эkonomyia toplyva y snyzhenye toksychnosty na avtomobylnom transporte. Transport, Moskva (1990)
26. Byrulia A (1964) Konstruirovaniye i raschet nezhestkikh dorozhnykh odezhd avtomobil'nykh dorog. Transport, Moskva
27. Lynnyk I (2017) Otsinka ta prohnozuvannia ekolohichnoho stanu dorozhnoho hospodarstva: monohrafiia. Kharkivs′kyy natsional′nyy universytet mis′koho hospodarstva im. O. M. Beketova, Kharkiv
28. Lynnyk I (1997) Optymizatsiia zakhodiv po znyzhenniu vydatkiv palyva ta vykydiv zaiu-rudniuiuchykh rechovyn avtomobiliamy pry rukhovi po miskym avtomobilnym mahistraliam. Kyiv

Chapter 3
Using the Maja Multi-criteria Method in Assessment of the Operation of Vehicles with Different Power Transmission Systems from the Perspective of Sustainable Urban Mobility

Ewelina Sendek-Matysiak and Grzegorz Sierpiński

3.1 Introduction

The turn of the 1960s and 1970s was marked with a breakthrough in thinking about natural environment and its connections with socio-economic development. As a result of such reflections, one could observe further new pieces of evidence that the concept of development oriented exclusively towards the growth of production, consumption and exploitation of nature had collapsed. The idea of new quality of life in greater harmony with nature, referred to as sustainable development, was becoming increasingly popular. The sustainable development concept was recognised as one of the cornerstones of development and policy making at the 1992 Earth Summit in Rio de Janeiro, where a global action plan for environmental protection, commonly referred to as Agenda 21, was adopted [1].

The demands of Agenda 21 were also reflected in the Treaty on European Union of 1992, while in the following years, sustainable development became one of the European Union's main pillars, which was ultimately reflected in the first Sustainable Development Strategy of 2001, officially entitled A Sustainable Europe for a Better World: A European Union Strategy for Sustainable Development [2]. With reference

E. Sendek-Matysiak (✉)
Department of Mechatronics and Machine Construction, Kielce University of Technology, Tysiąclecia Państwa Polskiego Av. 7, 25-314 Kielce, Poland
e-mail: esendek@tu.kielce.pl

G. Sierpiński
Faculty of Transport and Aviation Engineering, Silesian University of Technology, Krasińskiego Str. 8, 40-019 Katowice, Poland
e-mail: grzegorz.sierpinski@polsl.pl

E. Macioszek and G. Sierpiński (eds.), *Research Methods in Modern Urban Transportation Systems and Networks*, Lecture Notes in Networks and Systems 207, https://doi.org/10.1007/978-3-030-71708-7_3

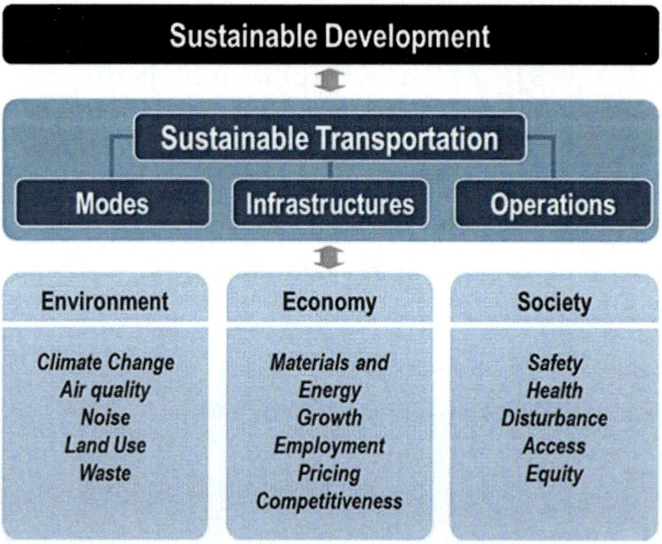

Fig. 3.1 Characteristics of sustainable transport development (based on [7])

to the definition coined by the World Commission on Environment and Development in 1987, in a report entitled Our Common Future (also known as The Brundtland Report), sustainable development perceived as a strategy component was defined as "development that meets the needs of the present without compromising the ability of future generations to meet their own needs" [3].

Transport is one of the sectors of economy which was highlighted as crucial for sustainable development from the very beginning. From the perspective of integrated order, i.e. the broad understanding of sustainable transport, implementation of transport as an activity means that passenger and cargo transportation is performed in a way that takes environmental, social and economic criteria into account at the same time (Fig. 3.1). It means that, among other aspects, it should be affordable, support the developing economy, offer a wide choice of the means of transport, reduce emissions and waste, minimise the use of non-renewable resources and land occupation, and reduce noise levels [4–6].

Sustainable transport has become the foundation for the development of the concept of sustainable urban mobility, whose importance grows particularly in light of the ongoing process of urbanisation (Fig. 3.2) and the deteriorating quality of urban life.

Sustainable urban mobility has also been one of the main priorities of the European Union transport policy for many years. In this context, one should refer to the 2007 Green Paper: Towards a new culture for urban mobility, which triggered an extensive discourse over the strategic importance of sustainable urban mobility for the entire Community [9].

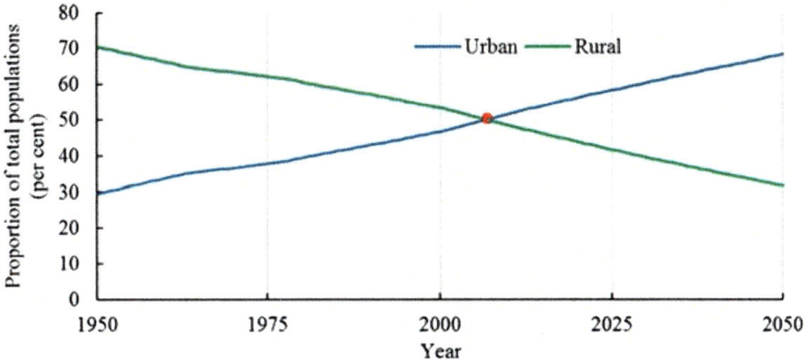

Fig. 3.2 Percentage of population in urban and rural areas of the world, 1940–2050 [8]

The vision of a competitive and sustainable transport system as well as the strategies enabling this goal to be fulfilled by 2030, with a 2050 perspective, were presented in the 2011 White Paper: Roadmap to a Single European Transport Area—Towards a competitive and resource efficient transport system [10]. At least two among the 10 strategic goals of the White Paper resonate with the new approach to the culture of mobility in cities and agglomerations, namely:

- reducing the number of conventionally-powered cars in urban transport by a half by 2030, eliminating them completely from cities by 2050, achieving CO_2-free logistics in large urban centres by 2030,
- by the year 2030, moving 30% of road freight transport over distances greater than 300 km to other modes of transport, e.g. rail or water transport, and by 2050—shifting more than 50% of this type of transport.

The final stage of the discussion initiated in 2007, adding more detail to the previous assumptions, was the adoption of the 2013 Urban Mobility Package whose implementation was intended to lead to creation of an urban transport system [11]:

- characterised by a high level of accessibility and catering to the mobility related needs of all users,
- functioning in accordance with the principles of sustainable development, introducing balance to the spheres of economic efficiency, social justice, health, quality of life, and environmental quality,
- optimised in terms of performance and cost-effectiveness, leading to more efficient use of space, and the existing infrastructure and services,
- making it possible to reduce the emission of noise, air pollutants, greenhouse gases, and energy consumption,
- contributing to the increasing attractiveness of the city as a safe place to live, and ultimately also to better functioning of the transport system throughout Europe.

Moreover, the document points out that due to the growing number of city dwellers in the EU (Fig. 3.3) and the preference for private car-based transport currently observed in the society [12, 13], the balancing of urban mobility should be founded

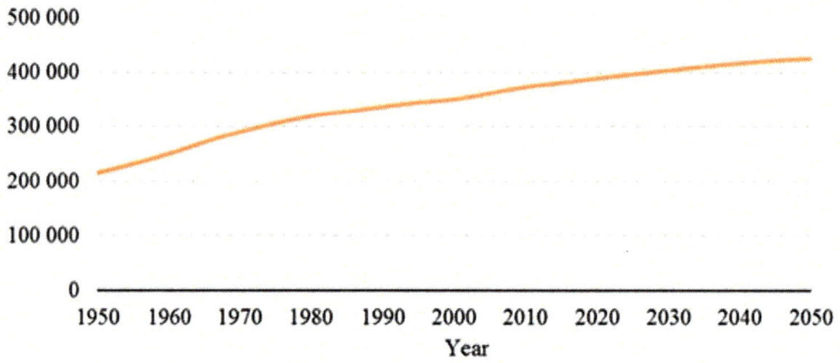

Fig. 3.3 Forecasted population of cities against the total population of the European Union (authors' own elaboration based on [14])

in the first place on placing vehicles running on alternative fuels, including electric energy from external power sources, on the market.

With regard to the foregoing, using one of the available methods of multi-criteria decision making support, a comparative analysis of cars powered by different energy sources has been provided further on in the paper, aimed to find an answer to the question of whether electrically driven cars will indeed contribute to the future development of sustainable urban traffic.

3.2 Using the Multi-criteria Decision Support Method for Assessment of Vehicles in the Urban Environment

It is doubtless that shaping a sustainable urban transport system is a complex decision-making process. In order to properly assess its development it is necessary to take many different perspectives into account, i.e. the relevant assessment criteria, which are not always consistent. The criteria applied to assess a project solution can be both quantitative and qualitative.

Typically, the solution assessment based on various criteria is expressed in different units (kilometres, tonnes, etc.). Therefore, in order to compare them, they should be expressed in the same (conventional) units, which one can achieve by normalising the assessment grades [15–17].

One of the normalisation methods used in the study addressed in this paper to assess different types and variants of cars is the Maja procedure [18–20]. The method's general capacity for being used when selecting the vehicle type that is most suitable for the given pre-set criteria has been discussed further on in this section by taking the environmental aspect into account. The latter is often omitted when analysing similar decision making problems. However, on account of the concern for the future and for the proper choice of actions that would conform with the sustainable development definition, more responsible planning is required, also in relation to vehicle fleets.

3.2.1 Procedure of the Maja Multi-criteria Assessment Method and Its Use for Vehicle Selection According to a Pre-set Criterion

Choosing the best variant by the Maja method makes it possible to use detailed assessment grades of vehicle variants by taking into account relative criteria importance factors. The solution boils down to calculating indicators of conformity and nonconformity of the assessment criteria, and using the dominance relationships to determine the non-dominated variant in the task at hand. Such an approach makes it possible to choose the best variant [21].

In general terms, the procedure of the Maja multi-criteria assessment method can be represented as follows:

1. Defining the ***V*** set of vehicle variants subject to the assessment.

$$V = \{v : v = 1, \ldots, N\} \tag{3.1}$$

where:

v—single vehicle variant,
N—number of vehicle variants ($N \geq 2$).

2. Defining the ***F*** set of assessment criteria with reference to which individual vehicle variants are assessed.

$$F = \{f : f = 1, \ldots, M\} \tag{3.2}$$

where:

f—single partial criterion,
M—number of partial criteria used in the assessment of the vehicle variants.

3. Preparing variant assessment matrices prior to normalisation **X**.

$$X = \left[x_{vf}\right]_{N \times M}, \quad v \in V, f \in F, x_{vf} \in R^{+} \tag{3.3}$$

where:

x_{vf}—assessment of the vth variant with regard to the fth partial criterion, known as diagnostic variables ($v \in V,\ f \in F$).

4. Normalising the variant assessment grades against individual criteria w_{vf} and creating matrix **W**, i.e.:

$$\mathbf{W} = \left[w_{vf}\right]_{N\mathbf{X}M'} \quad v \in V, f \in F, w_{vf} \in R^{+} \tag{3.4}$$

where:

$$w_{vf} = \begin{cases} \frac{x_{vf}}{\max_{v \in V} \cdot \{x_{vf}\}} & \text{for} \quad \text{stimulants} \\ \frac{\min_{x \in V} \cdot \{\mathbf{vf}\}}{x_{vf}} & \text{for} \quad \text{destimulants} \end{cases}$$

5. Assigning numerical values c_f representing the relative importance of the fth criterion to individual partial criteria comprising the $\boldsymbol{F}$ set, where $f \in F$. Typically, the more important the criterion, the higher its weight.
6. Creating the **Z** matrix of conformity.

$$\mathbf{Z} = [z_{vv}]_{N \times M}, \quad z_{vv}, \in [0, 1] \tag{3.5}$$

where:

z_{vv}—conformity index:

$$Z_{vv'} = \frac{1}{c} \sum_{f:f \in F:w_{vf} > w_{vf}} c_f, \quad c = \sum_{i=1}^{M} c_f \tag{3.6}$$

7. Creating the **N** matrix of nonconformity.

$$\mathbf{N} = [n_{v\sim}]_{N \times M}, \quad n_{vv} \in [0, 1] \tag{3.7}$$

where:

$$n_{vv^*} = \frac{1}{d} \max_{(v,f):w_{ivf} > w_{if}} \{w_{vf} - w_{vf}\} \tag{3.8}$$

d—difference between the element of the highest and that of the lowest value in the **W** matrix of normalised assessment grades

$$d = \max_{(v,f) \in V \times F} \{w_{vf}\} - \min_{(v,f) \in V \times F} \{w_{vf}\} \tag{3.9}$$

8. Determining threshold of conformity pz and threshold of nonconformity pn.
9. Developing the $\mathbf{A}^{\mathrm{b}}$ binary dominance matrix.

$$\mathrm{A}^b = [a_{vv}]_{N \times M} \tag{3.10}$$

where:

$$a_{vv} = \begin{cases} 1, & \text{when} \quad z_{vv} \geq pz \text{ i } n_{vv} \geq pn \\ 0, & \text{in other cases} \end{cases} \tag{3.11}$$

10. Developing the **Gf** graph of dominance.

$$\mathrm{Gf} = \langle Wf, Lf \rangle \tag{3.12}$$

for which:

Wf—set of vertices representing the set of analysed variants V,

Lf—set of arc (v, v') such that:

if $a_{vv'} = 1$, there is an arc from vertex v to vertex v';

if $a_{vv'} = 0$, there is no arc from vertex v to vertex v'.

The final choice of the vehicle variant has been made with reference to the **Gf** graph.

3.2.2 *Case Study*

The data used for the sake of the assessment of different car variants, i.e. those with petrol, diesel, plug-in hybrid and electric motors, concerned a single car model and came from the same manufacturer, i.e. Bayerische Motoren Werke AG. The territory of Poland was assumed as the area of analysis.

The vehicles confronted with one another represent the M1 passenger car category, characterised by identical or comparable total horsepower, the same type of body, power transmission system (front wheel drive), and gearbox (manual in cars with conventional engines and automatic in the electric ones).

The variants compared by the authors were as follows:

variant 1—car with a spark-ignition engine,

variant 2—car with a compression ignition engine,

variant 3—car with an electric engine,

variant 4—car with a hybrid Plug-In type drive system.

In terms of the assessment with reference to the basic parameters and the environmental impact of the cars tested, the following was taken into account:

1. technical performance factors:
   - maximum horsepower,
   - maximum torque,
   - acceleration to 100 km/h,
   - maximum speed,
   - average petrol consumption per 100 km,

- average diesel fuel consumption per 100 km,
- average energy consumption per 100 km,
- range (city);

2. economic factors:

(a) absolute measures:
- vehicle purchase cost,
- annual operating costs,

(b) relative measures:
- average energy consumption cost per 100 km,
- average fuel consumption cost per 100 km,
- vehicle maintenance cost per year;

3. qualitative factors:

- number of sales and service points,
- number of fuelling/charging stations,
- time required to replenish petrol/diesel fuel/electricity (when charging at an AC charging station),
- additional benefits, e.g. downtown parking, use of bus lanes, purchase allowances, etc.;

4. environmental (social) factors:

- CO_2 emission,
- total emission of CO, NO_x, HCs, PM10, SO_2, CO_2, CH_4, N_2O from the "fuel production" for the vehicle, for its operation and recycling/disposal,
- noise emission up to 50 km/h.

3.2.3 Results Analysis

The conclusion following an analysis of the graphs thus received and considering the least harmful impact on the urban environment (Fig. 3.4) is that the third variant,

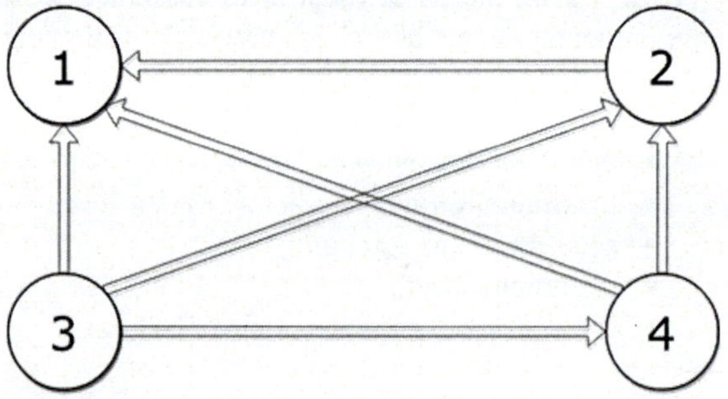

Fig. 3.4 Impact of vehicles with different power transmission systems on the urban environment—a graph of dominance

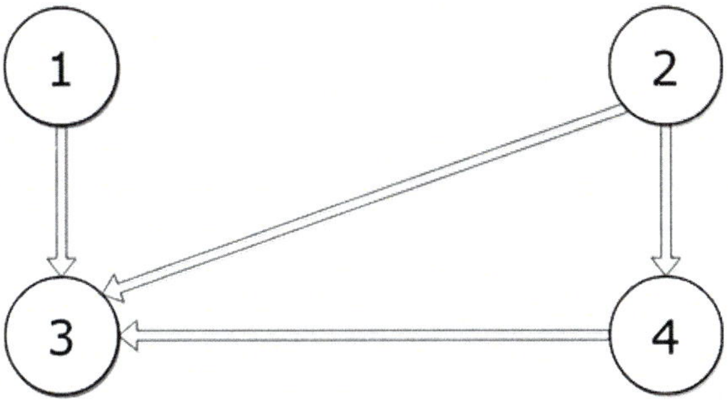

Fig. 3.5 Technical assessment of the vehicles subject to analysis

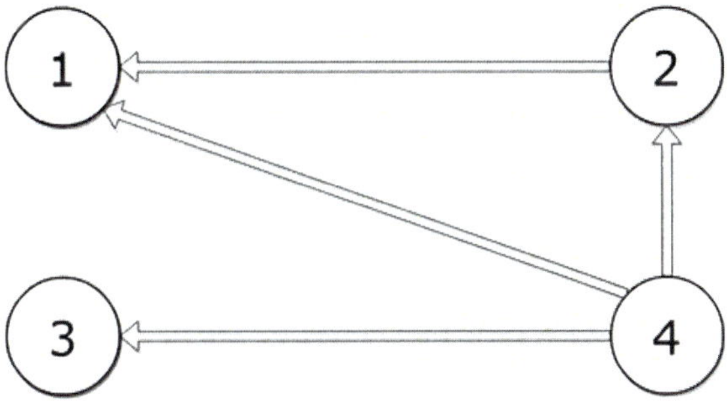

Fig. 3.6 Economic assessment of the vehicles subject to analysis

i.e. a car running exclusively on electricity, is the best one. Vertex no. 3 is the non-dominated one (with outgoing arcs only). On the other hand, the worst variant is number 1, i.e. a car with a petrol engine, since it features three incoming arcs and no outgoing ones (vertex 1 is dominated by vertices 2, 3 and 4).

From the technical point of view, the car equipped with the ZS engine (variant 2) turned out to be the best among the vehicles examined. Vertex no. 2 has only outgoing arcs (Fig. 3.5). The worst variant was the electric car (variant 3). Vertex 3 is dominated by vertices 1, 2 and 4. Based on the calculations, and taking the economic factors into account, the most advantageous option is using a hybrid car in the city. With this aspect in mind, the least positive alternative appears to be the vehicle with the ZI internal combustion engine (Fig. 3.6).

3.3 Conclusions

Adequate organisation of the urban transport system is particularly important for the proper functioning of cities, especially when considering the principles of sustainable development. One of the main problems of contemporary cities, and still unresolved, is individual transport, making it both possible and easier to move in a relatively unconstrained fashion, enabling comfortable access to different places, and improving the organisation of family and professional life. Since cars are perceived as particularly valuable goods, even determinants of social status, many people cannot imagine living without them. And it is precisely this mode of transport that has become the most crucial factor which shapes the spatial arrangement and determines the manner of functioning of cities as well as the mobility of their inhabitants [22].

In the document known as *Urban Mobility Package*, the European Commission encourages fundamental changes in the approach to both creating and managing urban mobility. At the same time, they have declared to further support actions and solutions that are favourable to sustainable urban transport, e.g. by substituting vehicles running on electricity for those with conventional engines, among other efforts. What has been demonstrated in this paper using the Maja multi-criteria assessment method is that, in contemporary conditions, electric cars exert a less harmful impact on urban environment than those with the internal combustion engine. Indeed, the advantage of electric vehicles used in urban areas is that not only do they not emit carbon monoxide, carbon dioxide, nitrogen oxides, and hydrocarbons where they are operated, but their noise emission is also very low, which is particularly desirable and important for shaping a sustainable urban transport system. However, from the technical point of view, such vehicles have received the worst ratings. Considering their limited range, the small number of charging stations, the long charging time, and the high cost of purchase, among other aspects, even despite their least deleterious impact on urban environment, the common interest they raise has remained very limited so far.

The case study provided in the paper was conducted for a particular country, namely Poland. Not until 2018 had the specific orientation of electromobility development been legally regulated in this country (under the law on electromobility and alternative fuels). And there has already been a significant increase of interest in electric vehicles (including the hybrid electric ones) observed ever since. Further charging stations are being commissioned in large cities, although their current number is still rather unimpressive compared to other countries. Therefore, the method used by the authors makes it also possible to perceive this problem in terms of assessment of the degree to which electromobility is developed in the chosen area.

The approach proposed in the paper may be included in the recommendations provided by municipalities to the operators of cars shared by users. Car sharing is in fact a service with a continuously growing stock of vehicles, and this trend is observed worldwide. Essentially, short-term rental of vehicles makes it possible to reduce the traffic volume in the road network of cities. And where the multi-criteria method is used to select the vehicle models for such a fleet, the system gains an additional advantage by minimising the negative environmental impact of transport.

References

1. UNCED (1992) Agenda 21. Rio declaration on environment and development. https://sustainabledevelopment.un.org
2. Miłaszewicz D, Ostapowicz B (2011) Warunki zrównoważonego rozwoju transportu w świetle dokumentów UE, „Gospodarka, Zarządzanie, Środowisko. Zeszyty Naukowe Uniwersytetu Szczecińskiego. Studia i Prace Wydziału Nauk Ekonomicznych i Zarządzania", nr 24, pp 103–118
3. European Commission (2001) Communication from the commission: a sustainable Europe for a better world: a European union strategy for sustainable development, Brussels 15.05.2001, COM(2001) 264 final
4. Motowidlak U (2017) Rozwój transportu a paradygmat zrównoważonego rozwoju. Studia Ekonomiczne. Zeszyty Naukowe Uniwersytetu Ekonomicznego w Katowicach, nr 337, pp 138–152
5. Esztergár-Kiss D, Braga Zagabria C (2020) Method development for workplaces using mobility plans to select suitable and sustainable measures. Research in Transportation Business & Management, online first. https://doi.org/10.1016/j.rtbm.2020.1005
6. Sierpiński G, Macioszek E (2020) Equalising the levels of electromobility implementation in cities. In: Mikulski J (ed) Research and the future of telematics, communications in computer and information science, vol 1289. Springer, Heidelberg, pp 165–176
7. Borys T (2009) Pomiar zrównoważonego rozwoju transportu [w:] D. Kiełczewski, B. Dobrzyńska (red.), Ekologiczne problemy zrównoważonego rozwoju, Wyd. Wyższej Szkoły Ekonomicznej w Białymstoku, Białystok, pp 166–185
8. United Nations. https://www.un.org/en/
9. European Commission (2007) Green paper–towards a new culture for urban mobility, COM(2007) 551 final, Brussels
10. European Commission (2011) White paper–roadmap to a single European transport area–towards a competitive resource efficient transport system, COM(2011) 144 final, Brussels
11. European Commission (2013) Annex 1. A concept for sustainable urban mobility plans to the communication from the commission to the European Parliament, the council, the European Economic and Social Committee and the Committee of the regions together towards competitive and resource-efficient urban mobility, COM(2013) 913 final, Brussels
12. Ministry of Infrastructure (2005) State transport policy 2006–2025, Warsaw
13. Statistical yearbook of the Republic of Poland from 2018 and previous years. https://stat.gov.pl/obszary-tematyczne/roczniki-statystyczne/
14. United Nations (2018) Departament of Economic and Social AFFairs: 2018 Revision of World Urbanization Prospects. https://www.un.org/development/desa/publications/2018-revision-of-world-urbanization-prospects.html
15. Sawicka H (2020) The methodology of solving stochastic multiple criteria ranking problems applied in transportation. Transp Res Procedia 47:219–226
16. Galińska B (2020) MCDM as the tool of intelligent decision making in transport. Case study analysis. In: Sierpiński G (ed) Smart and green solutions for transport systems. Advances in intelligent systems and computing, vol 1091. Springer, pp 67–79
17. Kijewska K, Iwan S, Małecki K (2019) Applying multi-criteria analysis of electrically powered vehicles implementation in Urban Freight Transport. Procedia Comput Sci 159:1558–1567
18. Ambroziak T, Lewczuk K (2009) Metoda wielokryterialna w zastosowaniu do oceny konfiguracji strefy składowania. Automatyka, T. 13, z. 2. Akademia Górniczo-Hutnicza im. Stanisława Staszica w Krakowie, pp 161–168
19. Goswami T, Pająk M, Skrzypiński W (2000) Biocompatibility of selected extractants in the continuous extractive ethanol fermentation. Inżynieria Chemiczna i Procesowa, T. 21, z. 4

20. Jacyna M (2001) Modelowanie wielokryterialne w zastosowaniu do oceny systemów transportowych. Prace Naukowe PW, seria Transport, z. 47, OWPW, Warszawa
21. Jacyna M, Turkowski D (2014) Wybrane aspekty wielokryterialnej oceny doboru środków transportowych w systemach dystrybucji pojazdów. Logistyka 4:1937–1946
22. Korneć R (2018) System transportu miejskiego wobec zrównoważonego rozwoju. Studia Miejskie 30:71–84

Chapter 4
The Role of Incentive Programs in Promoting the Purchase of Electric Cars—Review of Good Practices and Promoting Methods from the World

Elżbieta Macioszek

4.1 Introduction

When analyzing the state of advancement of electromobility in individual countries of the world, one can notice large differences. In Japan, the world's first electric vehicle without a battery was registered in 2016, which draws electricity from the road, and despite the drawbacks (such as low speed and the need to only move on a specially designed street that will be equipped with electrified tracks) it has an important positive feature—it is constantly powered and unlike electric cars with batteries, it will be able to cover any distance with the further development of this technology. In turn, China has become a tycoon in the sale of electric cars. In 2019, the number of newly registered electric cars increased by as much as 85% compared to 2018 [1]. In China, small-sized electric cars with a range of 100–250 km are most sold. In addition, the latest BYD SUV, the Song EV400 model, with a range of 350 km, which accelerates from 0 to 100 km/h in 4.8 s, is also very popular. The price of this type of vehicle is about $30,000 [1]. The European countries where the most electric cars are registered include: Norway, Germany, Great Britain and France (Fig. 4.1). In these countries, the demand for the purchase of an electric vehicle is strongly correlated with the proposed financial incentives.

The growing interest in electric vehicles is the result of the sustainable transport development policy, which, according to the assumptions presented in the White Paper, should be implemented by all EU countries. Currently, in Poland, compared to other Western European countries, the electric vehicle market is in the initial

E. Macioszek (✉)
Faculty of Transport and Aviation Engineering, Transport Systems and Traffic Engineering Department, Silesian University of Technology, Gliwice, Poland
e-mail: elzbieta.macioszek@polsl.pl

E. Macioszek and G. Sierpiński (eds.), *Research Methods in Modern Urban Transportation Systems and Networks*, Lecture Notes in Networks and Systems 207, https://doi.org/10.1007/978-3-030-71708-7_4

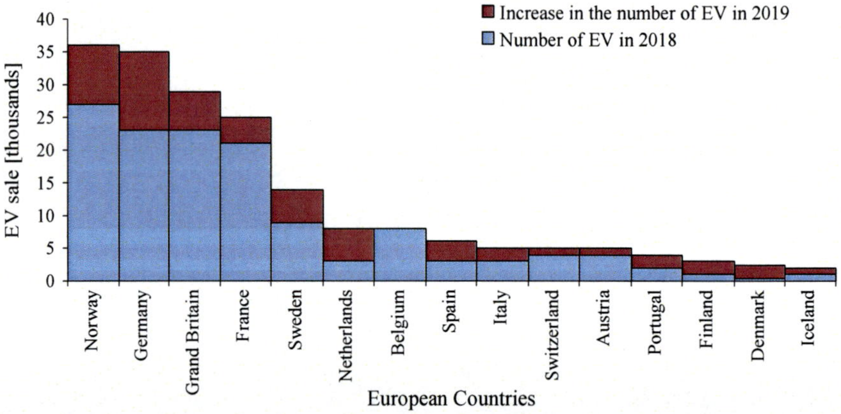

Fig. 4.1 The number of electric cars in Poland (*Source* Own based on data presented in [2])

stage of development, hence there are significantly fewer electric vehicles on Polish roads. The main reasons for the small number of electric cars in Poland are, first of all, relatively high prices of electric cars (Fig. 4.2), the lack or deficiencies in the

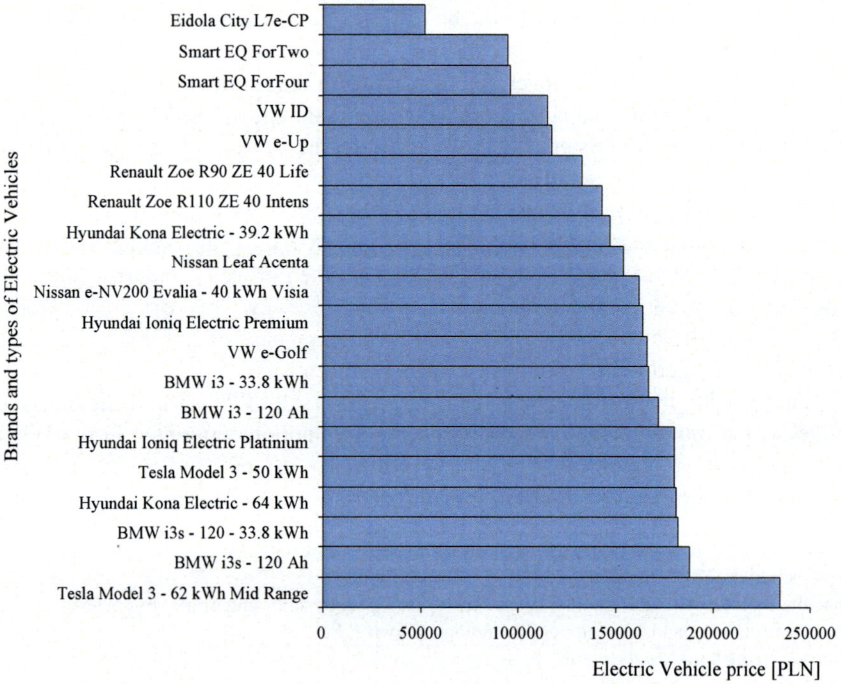

Fig. 4.2 The prices of electric cars in Poland (as of December 2019) (*Source* Own research based on data presented in [23])

available charging infrastructure, manifested by a small number of charging points for electric cars, and the technological problem related to the limited range of electric cars.

In Poland, in the footsteps of countries where electric vehicles are already a standard element of the vehicle fleet, we should expect a decline in the prices of electric cars and the development of charging infrastructure for electric vehicles in the near future. In recent years, many scientific papers devoted to electromobility and electric vehicles have been published (including: [3–21]). Pursuant to the provisions of the Act of January 11, 2018 on electromobility and alternative fuels (Journal of Laws 2018, item 317) [22], Polish cities have been obliged to prepare plans for the construction of electric vehicle charging stations. Pursuant to Art. 60.1. Of the Act, the minimum number of charging points to be installed by 31.12.2020 in publicly accessible charging stations located in municipalities is equal [22]:

- 1,000—in communes with a population greater than 1,000,000, where at least 600,000 motor vehicles have been registered and there are at least 700 motor vehicles per 1,000 inhabitants,
- 210—in communes with more than 300,000 inhabitants, where at least 200,000 motor vehicles have been registered and there are at least 500 motor vehicles per 1,000 inhabitants,
- 100—in communes with more than 150,000 inhabitants, where at least 95,000 motor vehicles have been registered and there are at least 400 motor vehicles per 1,000 inhabitants,
- 60—in communes with more than 100,000 inhabitants, where at least 60,000 motor vehicles have been registered and there are at least 400 motor vehicles per 1,000 inhabitants.

The chapter presents the issues of promoting electric cars based on incentive programs, preliminary results of research on the role and importance of incentive programs in promoting the purchase of electric cars in Poland, and a review of the so-called good practices from around the world in the field of promoting electric cars and the operation of electric car charging points. The conclusions drawn from these practices can be a valuable source of information, tips and guidelines when carrying out works at the initial stage of electromobility development in Poland.

4.2 The Electric Car Promotion Policy Based in Incentive Programs

Many countries in the world, in order to overcome the obstacles related to the implementation of electric cars in everyday life, have developed a policy of promoting electric cars based on various incentive programs (so-called incentives). Several classifications of incentives can be found in the literature on the subject. One of the classifications generally distinguishes three main groups of incentives. There are:

- incentives related to the producer of electric cars,
- incentives related to the charging infrastructure for electric cars,
- financial incentives aimed directly at owners of electric cars or owners of electric car fleets.

The forms of incentive programs used when purchasing an electric car presented in various scientific papers are presented in Table 4.1.

In turn, Table 4.2 presents the results of the analysis of incentives to purchase electric cars used in various European countries, based on a review of the literature on the subject (including: [24–33]). On the basis of the obtained results, it can be concluded that the level of applying incentives in individual European countries differs significantly. Norway and England are among the countries with the highest number of financial incentives. In some countries such as France, Ireland, several financial incentives are used when purchasing electric cars. On the other hand, Poland is a country where so far no incentive programs dedicated to electric vehicle owners have been implemented (apart from concessions at the local government level, which belong to the group of the so-called soft support instruments). In some Polish cities, owners of electric cars have been entitled to enter the zone of privileged, zero or reduced parking fees. However, there are no financial incentives in Poland.

An interesting form of financial incentive is the scrapping program. The main objective of the program is to improve the air quality in a given country by scrapping older vehicles. Owners of older diesel cars who decide to scrap their car are compensated in cash for the purchase of vehicles with lower emissions. So far, such incentive programs have been introduced in Germany and Great Britain. In Great Britain, the scrapping program is carried out by the PSA group [34]. The program is dedicated to owners of various vehicle brands. The car must be registered on the person interested in the offer for at least 90 days.

As stated in the research work [34], the maximum amount of financing for the purchase of a new vehicle in the amount of GBP 7,000 can be obtained. The amount of the grant depends on the vehicle model. This type of action is related to the promotion of cars that meet the latest emission standards. Each car that will be returned to the showrooms of three French brands (Peugeot, Citroen, DS) will be scrapped and recycled.

In all European countries listed in Table 4.2, taxes on fossil fuels have been increased along with the promotion and development of electromobility. This fact, in a sense, acts as an indirect incentive program.

4.3 The Review of Good Practices, Methods in the Promotion of Electric Cars in the World

In order to promote electromobility, the governments of many countries around the world have prepared various types of incentive programs for electric car users. The leader in terms of the incentives offered is undoubtedly Scandinavia (especially

Table 4.1 The forms of incentive programs used when purchasing an electric car

Author(s)	Incentive program	Characteristic
Kley et al. [35]	Legal regulations	By using legal provisions, various types of restrictions can be imposed on car manufacturers, e.g. specific emission levels for newly manufactured vehicles. These limits focus on inputs, outputs or define certain required production processes (e.g. production standards, mandatory emission targets)
	Economic incentives	With the use of economic incentives, market performance can be shaped by price or quantity changes (e.g. CO_2 certificates, tax cuts, subsidies, scrapping the program, congestion charges, parking fees)
	Use of suggestive means	Used to persuade buyers and producers by providing information, creating better administrative conditions, financing research and development programs (eg. information campaigns, developing standards (eg. plugins))
	Organizational incentives	By using organizational incentives, it is possible to reduce obstacles, e.g. development of the necessary infrastructure, the appointment of supervisory authorities to control market structures (e.g. charging infrastructure, free parking spaces)
Lingzhi et al. [36]	Direct incentives	Direct incentives are associated with a direct financial benefit for the consumer (e.g. purchase subsidies, tax reductions, subsidies or full financing of electric vehicle equipment, exemption from charges for electricity, free parking, exemptions from emission tests)
	Indirect incentives	Consumers do not benefit directly from the use of indirect incentives, but this type of incentive saves time and provides convenience. Examples: exemption from emissions testing, access to public vehicle charging stations, time savings

(continued)

Table 4.1 (continued)

Author(s)	Incentive program	Characteristic
	Discouraging factors	Schemes, rules, regulations, guidelines, etc. have an unintended negative impact on the attractiveness of electric vehicles (e.g. a mandatory annual fee for using an electric car to compensate for lost tax revenues from the purchase of petroleum fuels)
	Other types of incentives	Other approaches to developing the electric vehicle market (e.g. zero-emission schemes, research and development, rebates and insurance protections, incentives offered by utilities or incentives for the electric vehicle fleet)
Laurent and Windisch [37]	Economic incentives	Actions aimed at overcoming the price barrier related to the high cost of purchasing electric vehicles, supporting the development of new technologies in the field of electric vehicles, providing financial incentives to potential buyers (e.g. direct investments in research and development and infrastructure, preferential pricing policy, subsidies for the purchase of electric vehicles or subsidies for the construction of new charging point infrastructure, tax incentives for owners of electric vehicles)

Norway, where the share of electric cars in the automotive market is constantly growing). On a global scale, a wide range of incentive programs is mainly offered by countries such as China, Japan and the USA. This subchapter presents an overview of the electric car incentive program offered by different countries. These programs can be considered as examples of good practice in promoting electric cars.

In Norway, the rapid change of the fleet to electric vehicles is supported by low prices of electricity obtained from hydroelectric power plants (which cover 99% of the country's electricity needs) and the wealth of Norwegians, who are among the richest societies in the world. In terms of transport policy, there are many financial incentives for people buying electric cars. Electric car owners in Oslo [38–40]:

- are exempt from the vehicle purchase tax when purchasing an electric vehicle and are exempt from VAT,
- are entitled to a price reduction for the registration of a new electric car,
- use specially designated parking spaces in city centres,

Table 4.2 The analysis of incentive programs used in different countries in Europe

Type of incentive program	Poland	England	Germany	Belgium	Croatia	France	Sweden	Norway	Finland	Ireland	Austria	Switzerland	Denmark	Spain	Italy	Netherlands	Greece
Exemption from parking fees	–	+	–	–	–	–	–	+	–	–	–	–	–	–	+	–	–
Exemption from the tourist tax	–	+	–	–	–	–	–	+	–	–	–	–	–	–	+	–	–
Discounts on annual taxation of eclectic vehicles	–	+	+	+	+	–	+	+	–	–	+	+	+	+	–	+	+
Different types of subsidies for electric vehicle owners	–	+	–	–	–	+	–	–	–	–	–	–	–	+	+	–	–
Tax reduction when buying an electric vehicle	–	–	–	+	–	–	–	+	–	+	+	+	+	+	–	+	+

(continued)

Table 4.2 (continued)

Type of incentive program	Poland	England	Germany	Belgium	Croatia	France	Sweden	Norway	Finland	Ireland	Austria	Switzerland	Denmark	Spain	Italy	Netherlands	Greece
Tax reduction after buying an electric vehicle ara>	–	–	–	+	–	–	–	–	–	–	–	–	–	–	–	–	–
Scrapping program	–	+	+	––	–	–	–	–	–	–	–	–	–	–	–	–	–

- are exempt from parking fees,
- are exempt from fees for the use of roads, bridges and ferries,
- have a permit to use bus lanes.

On the basis of the research presented in [24], can be stated that the greatest role among the incentives was played by the tax exemption and the VAT exemption when purchasing an electric car. As a result of these incentives, the number of registered electric cars in 2017 increased to over 200,000, and thus the total sales of hybrid and electric cars exceeded the sales of combustion-powered cars.

In turn, China is the world's largest market for electric cars. As reported in [41], 40% of the 2,000,000 electric cars circulating around the world are in China. 94% of the electric car market in China is owned by local companies. According to forecasts, from 2020, 12% of electric cars sold are to be electrically powered. In 2018, there were 487 electric car manufacturers in China. The policy of incentives in Chinese provinces is as follows [42, 43]:

- organization of numerous demonstration programs—consisting in making electric cars available to the local population for test drives,
- subsidy for the purchase of electric cars used for public purposes, i.e. long-distance buses, taxis and subsidies for buyers of private electric cars (the amount of the subsidy depends on the capacity of the battery and the type of electric vehicle),
- exemption from fees for registering an electric car,
- free charger when buying an eclectic car,
- subsidy for charging up to 3 years or up to 60,000 km,
- cash for converting conventional cars to electric cars,
- exemption from road tolls.

In Japan, in 2009, the government published a new national energy strategy to 2030, which aims to achieve two main goals:

- improvement of the fuel efficiency of new vehicles by 30 %,
- reducing dependence in powering vehicles on petroleum-derived fuels.

The incentive policy in Japanese prefectures is as follows:

- subsidies for the purchase of new electric cars for individuals,
- exemption from tax on the purchase of an electric car,
- toll discounts,
- free parking spaces.

Granting the above-mentioned incentives vary by prefecture.

In turn, in the USA, the "American Recovery and Reinvestment Act" of 2012 [44] assumed the allocation of USD 800 billion to create new jobs and economic growth of the country through investments in renewable energy and improvement of energy efficiency, including tax credits for the purchase of electric cars. The policy of incentives in the USA is as follows [29, 42, 45, 46]:

- subsidies for the purchase of new electric cars for individuals (the amount of the subsidy depends on the battery capacity),
- subsidies for the purchase of a charger,
- lower electricity rates for people who have a separate meter for charging electric cars (in some US states),
- 18 US states offer tax financial incentives, including electric car sales tax exemption, fuel tax exemption, reduced license and/or electric car use tax, exemption or reduction of the cost of installing an electric car charger,
- reduced car registration fee,
- special, advantageously interest-bearing loans for the conversion of a conventional vehicle into an electric car,
- exemption from vehicle emission control tests,
- exemption from parking fees,
- reduced rates of tolls for toll roads.

Above mentioned incentives vary from state to US state. Due to numerous incentive programs, electric cars dominate in large metropolitan areas.

4.4 The Preliminary Study of the Role of Incentive Programs in Promoting the Purchase of Electric Cars under Polish Conditions

A research survey was conducted in order to initially understand the role of incentive programs in promoting the purchase of electric cars. The research was conducted on-line using the "ProfiTest" web application in 2017 and 2018. The survey was addressed to Polish residents. During the survey, 1,814 respondents started answering, of which 1,777 completed the questionnaire correctly and completely. The characteristics of the research sample are presented in Table 4.3. Due to the method of selecting the respondents, the conducted surveys were quasi-representative research, as the surveyed group of respondents met the requirements of the representative method only in some respects. In the initial part of the questionnaire, a short theoretical introduction was provided, which was aimed at introducing the issues related to the broadly understood electromobility, infrastructure related to charging electric vehicles, including e.g. definitions of individual types of electric vehicles, classifications of plugs, points, electric car charging stations, etc. The next part of the questionnaire concerned general questions allowing to characterize the respondents. In the main part of the questionnaire, the respondents assessed the importance of incentive programs to buy an electric car. This part of the survey consisted of specific questions relating to specific incentives. In the questions in which the respondents were asked to organize their answers in an ascending or descending order, or to indicate the most favourable incentive, the ranking was used. In all questions, the best—according to the respondents—the solution was given a rank of 1.00, while the worst—a rank of 6.00. The mean rank value was then calculated. The survey

Table 4.3 The characteristics of the research sample ($n = 1777$)

Feature		Structure of respondents
Gender	Men	59%
	Woman	41%
Age	18–25 years	6
	26–35 years	245
	36–45 years	546
	46–55 years	653
	>55 years	327
Education	Basic	3
	Medium	846
	Higher	928
Monthly net income	<2500	7
	2501–3000	101
	3001–3500	143
	3501–4000	292
	4001–4500	515
	4501–5000	480
	5001–5500	211
	>5500	28
Average number of cars per household		1.12
Area of Poland	Silesian—Zagłębie Conurbation	690
	Warsaw Agglomeration	583
	Northeast	8
	Northwest	370
	Southeast	22
	Southwest	104

contained single and multiple-choice closed questions as well as ranking questions. Due to the limitations of the volume of the chapter, selected results of the survey are presented below.

Based on the data presented in Table 4.3, it can be concluded that the studied sample is dominated by men (59%), the 46–55 age group, people with secondary and higher education, with an average income of PLN 4,000–5,000 gross, living in the Silesian and in the Warsaw agglomeration, where awareness of problems related to transport and environmental protection is particularly high. Six different incentives were analyzed in the study. They were:

- prices of electric cars comparable to the prices of cars powered by petroleum fuels,
- VAT exemption when buying an electric car,
- exemption from road tolls (e.g. from tolls on toll road sections),
- exemption from parking fees,
- possibility of using bus lanes,
- access to a dense network of charging points for electric cars.

An important incentive was also considered to be the reduction or complete exemption from the charge for charging the vehicle at publicly accessible charging points for electric cars. Nevertheless, in Poland, the charging infrastructure is still in the initial stage of development, therefore it was considered that the more important problem than the price is the availability and dense network of charging points for electric cars. The results of the analysis of incentive programs when purchasing an electric car are presented in Fig. 4.3.

When analyzing the data presented in Fig. 4.3, it can be concluded that the financial aspect, related to the price of an electric car, is undoubtedly the most important for all respondents. The second place in the opinion of respondents is the availability of electric car charging points. The significance of incentive programs was also analyzed, broken down into homogeneous groups of incentives related to:

- reducing the cost of purchasing electric cars,
- reduction of costs of using electric cars,
- infrastructure incentives.

Based on the analysis of the respondents' answers, it can be concluded that in the group of incentives related to the reduction of the purchase costs of electric cars, the respondents again recognized the price of the vehicle as to the most important feature (99%). In the group of incentives related to the reduction of costs of using electric cars, the respondents indicated exemption from road tolls (73%), and then exemption from parking fees (27%). On the other hand, in the group of incentives related to infrastructure, access to charging points turned out to be the most important

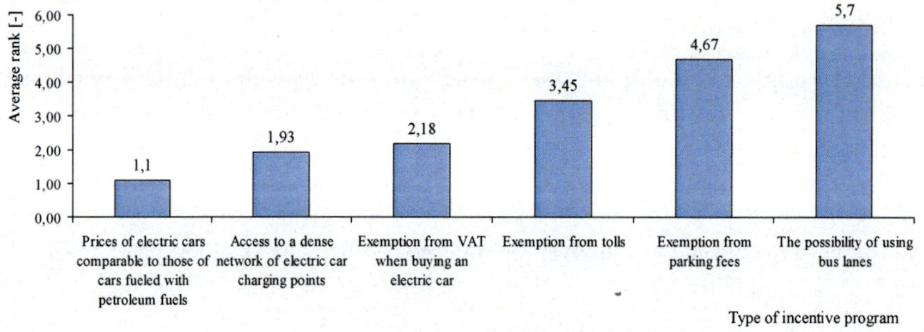

Fig. 4.3 The results of the analysis of the significance of incentive programs when purchasing an electric car

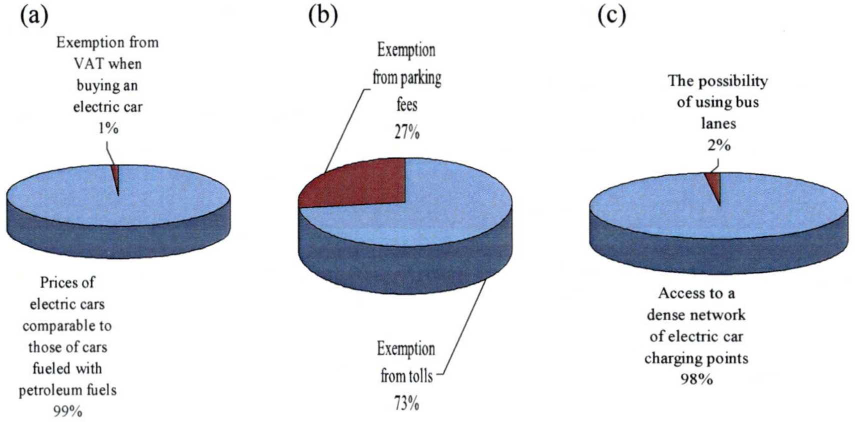

Fig. 4.4 The results of the analysis of the significance of incentive programs when buying an electric car, broken down into homogeneous groups of incentives

(98%). This fact is reflected in reality, as Poland does not yet have a dense network of charging points. The least important feature when using an electric car was considered to be the possibility of using bus lanes (2%). The results of the analysis of the value of incentive programs when buying an electric car broken down into homogeneous groups of incentives are presented in Fig. 4.4.

The obtained data also allowed for the analysis of the significance of incentive programs when purchasing an electric car, taking into account such characteristics as gender, age, net monthly income and the respondent's place of residence. The results of the analyzes are presented in Figs. 4.5 and 4.6. Based on the analyzes carried out in this area, it can be concluded that:

- regardless of the analyzed feature, the most important incentive program was the purchase price of an electric car, and the least important—the possibility of using bus lanes,
- for men, the availability of a dense network of charging points for electric cars turned out to be more important than for women,
- residents of south-eastern Poland indicated the availability of a dense network of electric car charging points as the least important feature.

4.5 Conclusions

According to forecasts, a significant increase in the number of electric vehicles on Polish roads should be expected in the next years. The use of electric drive in the vehicle results in the reduction of dust and gas emissions, harmful both to health and the environment, and reduces the emission of CO_2 into the atmosphere, eliminating the combustion process of hydrocarbons, i.e. gasoline and diesel oil. Hence, the

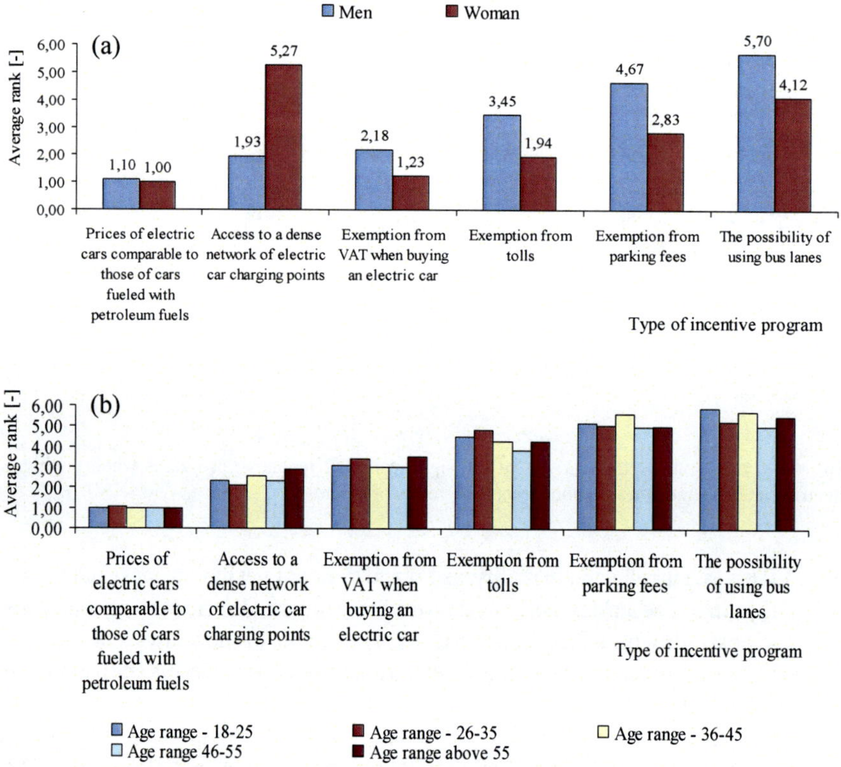

Fig. 4.5 The results of the anaalysis of the significance of incentive programs when buying an electric car, **a** taking into account respondent gender, **b** respondent age

extensive introduction of electric cars and buses onto Polish roads may be of key importance for air purity in selected areas of the country.

Electric cars (with the exception of Norway) tend to be more expensive to buy than petroleum-based vehicles. International literature [47, 48] indicates that the economic status of the first buyers of electric cars in a given country is much higher than the average. This situation is also reflected in Poland.

The analysis of the role of incentive programs in promoting the purchase of electric cars carried out in the chapter allows for the conclusion that in Europe, electric cars are the most popular in the Nordic countries, i.e. in Norway, Sweden, Finland and Iceland. Denmark is an exception, but this is mainly due to the lack of incentive programs dedicated to electric car owners. This translates into the number of electric car owners. Moreover, when analyzing the charging points themselves (there are about 100,000 of them in the EU), it should be stated that the disproportions in their number are very large. The vast majority (76%) are located in only four EU countries, i.e. the Netherlands (28%), Germany (22%), France (14%) and the United Kingdom (12%). The lowest in Cyprus and Greece (about 40 units). According to

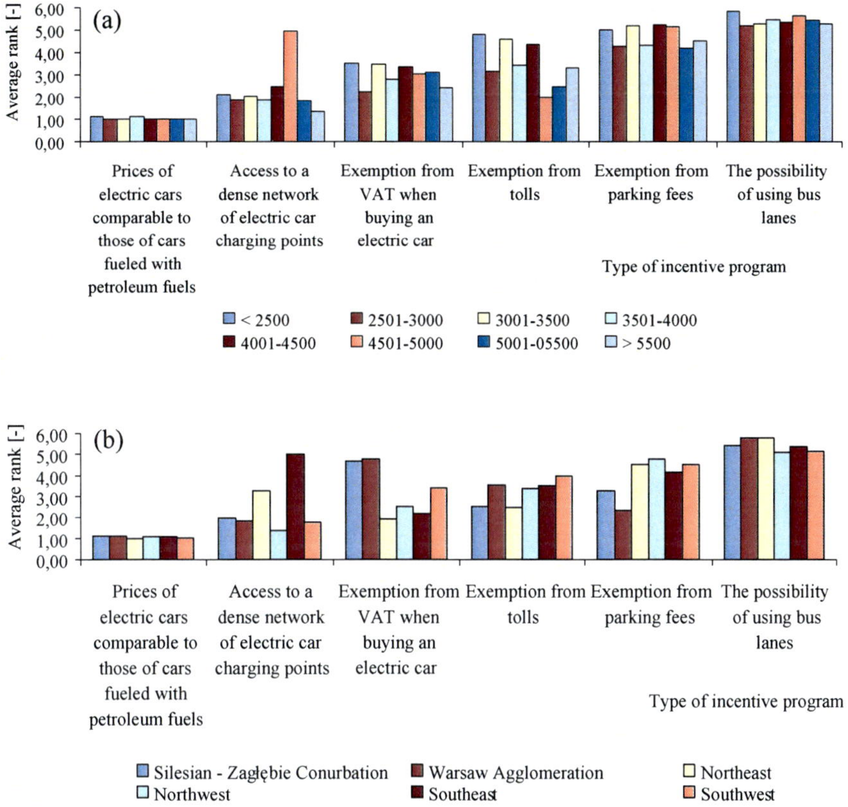

Fig. 4.6 The results of the analysis of the significance of incentive programs when buying an electric car, taking into account **a** monthly net income (PLN), **b** the place of residence of the respondents

European Automobile Manufacturers Association [49], if the electric car fleet were to be electrified by 2025 at the rate expected by the EU, approximately 2,000,000 recharging points would be needed.

The results of preliminary research on the role of incentive programs in promoting the purchase of electric cars in Poland allow for the conclusion that undoubtedly the most important aspect for all respondents is the financial aspect related to the price of an electric car. The second place in the opinion of respondents is the availability of charging points for electric cars. The least significant incentive was the possibility of using bus lanes. Due to the fact that in Poland the issue of electro-mobility is at the initial stage of development, the presented research takes the form of pilot studies that should be continued in the future.

References

1. Błoński M, Chiny potentatem sprzedaży samochodów elektrycznych, https://kopalniawiedzy.pl/Chiny-samochod-elektryczny-BYD-Tesla, 28114
2. EV Volumes.com. The electric vehicle world sales database, https://www.ev-volumes.com
3. Ambroziak T, Gołębiowski P, Jacyna-Gołda I, Kłodawski M, Pyza D (2015) Ocena parametrów wpływających na zwiększenie poziomu zanieczyszczeń emitowanych przez pojazdy na przykładzie niektórych obszarów. Logistyka 6:1400–1409
4. Bakun B (2015) Zagadnienie jednostkowego zużycia energii pojazdu elektrycznego na przykładzie samochodu osobowego po konwersji napędu. Przegląd Elektrotechniczny 91(2):174–176
5. Chłopek Z (2012) Badanie zużycia energii przez samochód elektryczny. Pimot 57:105–117
6. Gis W, Menes M (2015) Stan i perspektywy rozwoju samochodów elektrycznych. Logistyka 4:3393–3399
7. Gis W, Menes E, Waśkiewicz J (2016) Elektromobilność i wodoryzacja transportu samochodowego w Polsce. Logistyka 44:83–94
8. Krzak J (2012) Ograniczanie emisyjności pojazdów w transporcie drogowym. Pojazdy hybrydowe i elektryczne w Polsce—perspektywy i bariery rozwoju. Studia Biura Analiz Sejmowych 1(29):161–184
9. Macioszek E (2019) Freight transport planners as information elements in the last mile logistics. In: Sierpiński G (ed) integration as solution for advanced smart urban transport Systems. Advances in Intelligent Systems and Computing 844. Springer International Publishing Switzerland, pp 242–251
10. Macioszek E.: First and last mile delivery—problems and issues. [in:] G. Sierpiński (ed.) Advanced Solutions of Transport Systems for Growing Mobility. Advances in Intelligent Systems and Computing 631. Springer International Publishing Switzerland, pp. 147–154 (2018)
11. Macioszek E (2019) E-mobility infrastructure in the Górnośląsko—Zagłębiowska Metropolis, Poland, and potential for development. Proceedings of the 5th World Congress on New Technologies (NewTech'19), Lisbon, Portugal, August 18–20, 2019. Paper No. ICERT 108, pp. ICERT 108-1-ICERT 108-4, https://doi.org/10.11159/icert19.108.
12. Merkisz J, Pielecha I (2006) Alternatywne napędy pojazdów. Wydawnictwo Politechniki Poznańskiej, Poznań
13. Merkisz J, Pielecha I (2015) Układy elektryczne pojazdów hybrydowych. Wydawnictwo Politechniki Poznańskiej, Poznań
14. Molecki A (2015) Elektryczne Środki Komunikacji Miejskiej. Autobusy 9:18–22
15. Murawski J, Szczepański E (2014) Perspektywy Rozwoju Elektromobilności W Polsce. Logistyka 4:2249–2258
16. Wójtowicz S (2011) Pojazdy elektryczne i sieci Smart Grid. Instytut Elektrotechniki, Warszawa
17. Flasza J, Matuszczyk P (2018) Elektromobilność w Polsce a systemy OZE. Przegląd Elektrotechniczny 1:33–36
18. Pyza D, Gołaszewski A, Jacyna M, Gołębiowski P (2017) Proekologiczny system transportowy w aspekcie rozwoju infrastruktury paliw alternatywnych. Prace Naukowe Politechniki Warszawskiej. Seria Transport 119:355–366
19. Sendek-Matysiak E, Pyza D (2018) The assignment of vehicle assessment based on multi criteria decision making. Arch Transp 48(4):77–85
20. Macioszek E (2019) Changes in values of traffic volume—case study based on general traffic measurements in opolskie Voivodeship (Poland). In: Macioszek E, Sierpiński G (eds) Directions of development of transport networks and traffic engineering. Lecture Notes in Networks and Systems, vol 51, pp 66–76. Springer International Publishing Switzerland
21. Macioszek E (2020) Electric vehicles—problems and issues. In: Sierpiński G (ed) smart and green solutions for transport systems. Advances in Intelligent Systems and Computing, vol 1091, pp 169–183. Springer Nature Switzerland AG (2020)

22. Kancelaria Sejmu: Ustawa z dnia 11 stycznia 2018 r. o elektromobilności i paliwach alternatywnych (Dz.U.2018 poz. 317)
23. Electric car prices in Poland, https://elektrowoz.pl/auta/aktualne-ceny-samochodow-elektrycznych-w-polsce-pazdziernik-2018/
24. Bjerkan KY, Norbech TE, Nordtomme ME (2016) Incentives for promoting Battery Electric Vehicle (BEV) adoption in Norway. Transp Res Part D 43:169–180
25. Figenbaum E (2017) Perspectives on Norway's supercharged electric vehicle policy. Environmental Innovation and Societal Transitions 25:14–34
26. Hardman S, Shiu E, Steinberger-Wilckens R (2016) Comparing high-end and low-end early adopters of battery electric vehicles. Transp Res Part a 88:40–57
27. Harrison G, Thiel C (2017) An exploratory policy analysis of electric vehicle sales competition and sensitivity to infrastructure in Europe. Technol Forecast Soc Charge 114:165–178
28. Jacyna M, Lewczuk K, Szczepański E, Gołebiowski P, Jachimowski R, Kłodawski M, Pyza D, Olena S, Wasiak M, Żak J, Jacyna-Gołda I (2015) Effectiveness of national transport system according to costs of emission of pollutants. In: Nowakowski T (ed.) Safety and reliability: methodology and applications, pp 559–567. CRC Press Taylor & Francis Group
29. Jin L, Searle S, Lutsey N (2014) Evaluation of state-level U.S. Electric vehicle incentives. White paper. The International Council on Clean Transportation. European Commission, Brussel (2014)
30. Krupa JS, Rizzo DM, Eppstein MJ, Lanute DB, Gaalema DE, Lakkaraju K, Warrender ChE (2014) Analysis of a consumer survey on plug-in hybrid electric vehicles. Transp Res Part a 64:14–31
31. Levay PZ, Drossinos Y, Thiel C (2017) The effect of fiscal incentives on market penetration of electric vehicles: A pairwise comparison of total cost of ownership. Energy Policy 105:524–533
32. Rezvani Z, Jansson J, Bodin J (2015) Advances in consumer electric vehicle adoption research: A review and research agenda. Transp Res Part D 34:122–136
33. Zhou Y, Wang M, Hao H, Johnson L, Wang H (2015) Plug-in electric vehicle market penetration and incentives: a global review. Mitig Adapt Strat Glob Change 20(5):777–795
34. A proposal from the PSA Group in response to the scrapping program for non-organic cars in Great Britain, https://www.francuskie.pl/grupa-psa-program-zlomowania/
35. Kley F, Wietschel M, Dallinger D (2010) Evaluation of European Electric Vehicle Support Schemes. Fraunhofer Institute for Systems and Innovation Research ISI. Frauhofer, Karlsruhe
36. Lingzhi J, Searle S, Lutsey N (2014) Evaluation of state-level U.S. electric vehicles incentives. White paper. The International Council on Clean Transportation, Washington
37. Laurent F, Windisch E (2011) Triggering the development of electric mobility: a review of public policies. Europ Transp Res Rev 3:221–235
38. Aasness MA, Odeck J (2015) The increase of electric vehicle usage in Norway—incentives and adverse effects. Europ Transp Res Rev 7:34. Springer, Berlin Heidelberg (2015)
39. Haugneland P, Kvisle HH (2013) Norwegian electric car user experiences. 2013 World Electric Vehicle Symposium and Exhibition 2013, https://www.evs28.org/event_file/event_file/1/pfile/Haugneland=Hauge_norwegian-Elcetric-Car-User-Experiences-2014.pdf
40. Mersky AC, Sprei F, Samaras C, Qian ZS (2016) Effectiveness of incentives on electric vehicle adoption in Norway. Transportation Research Part D: Transport and Environment 46:56–68
41. Motoryzacja, https://motoryzacja.wnp.pl/chiny-sa-najwiekszym-na-swiecie-rynkiem-samochodow-elektrycznych,307878_1_0_0.html
42. Helveston JP, Liu Y, Feit EMcD, Fuchs E, Klampfl E, Michalek JJ (2015) Will subsidies drive electric vehicle adoption? Measuring consumer preferences in the U.S. and China. Transp Res Part A: Policy Pract 73:96–112
43. Song Y, Yang X, Lu Z (2010) Integration of plug-in hybrid and electric vehicles: Experience from China. IEEE PES General Meeting, Providence, RI, pp 1–6
44. American Recovery and Reinvestment Act, https://www.thebalance.com/arra-details-3306299
45. Carley S, Krause RM, Lane BW, Graham JD (2013) Intent to purchase a plug-in electric vehicle: a survey of early impressions in large US cities. Transp Res Part D: Transp Environ 18:39–45

46. Diamond D (2009) The impact of government incentives for hybrid-electric vehicles: Evidence from US states. Energy Policy 37(3):972–983
47. Campbell AR, Ryley T, Thring RH (2012) Identifying the early adopters of alternative fuel vehicles: a case study of Birmingham United Kingdom. Transp Res Part A 46:1318–1327
48. Plotz P, Schneider U, Globisch J, Dutschke E (2014) Who will buy electric vehicles? Identifying early adopters in Germany. Transp Res A: Policy Pract 67:96–109
49. European Automobile Manufacturers Association, https://www.acea.be/

Chapter 5
Life-Cycle Costing Decision-Making Methodology and Urban Intersection Design: Modelling and Analysis for a Circular City

Orazio Giuffrè, Anna Granà, Tullio Giuffrè, Francesco Acuto, and Anthony Lo Pinto

5.1 Introduction

Road infrastructures evolved over time to be increasingly efficient and safe, but their construction, maintenance, operation and renewal often proved to be costly for user communities and damaging to the environment. To meet the economic, environmental and social needs of the future mobility, research and innovation in transportation are shifting from optimizing only functional targets toward sustainability and modernization aspects in each phase of the life-cycle of the transport infrastructures [1]. Transportation agencies are now increasingly looking at costs and impacts in all phases of the life-cycle of the road infrastructure projects when strategic decisions and/or individual investment should be made [2].

In recent years, life-cycle thinking-based approaches have developed and spread progressively to ensure sustainability of products, designs, services, technologies and systems [3, 4]. The life-cycle thinking is the very core of circular economy that

O. Giuffrè · A. Granà (✉) · F. Acuto · A. Lo Pinto
Department of Engineering, University of Palermo, Palermo, Italy
e-mail: anna.grana@unipa.it

O. Giuffrè
e-mail: orazio.giuffre@unipa.it

F. Acuto
e-mail: francesco.acuto@unipa.it

A. Lo Pinto
e-mail: anthony.lp92@gmail.com

T. Giuffrè (✉)
University of Enna Kore, Enna, Italy
e-mail: tullio.giuffre@unikore.it

E. Macioszek and G. Sierpiński (eds.), *Research Methods in Modern Urban Transportation Systems and Networks*, Lecture Notes in Networks and Systems 207, https://doi.org/10.1007/978-3-030-71708-7_5

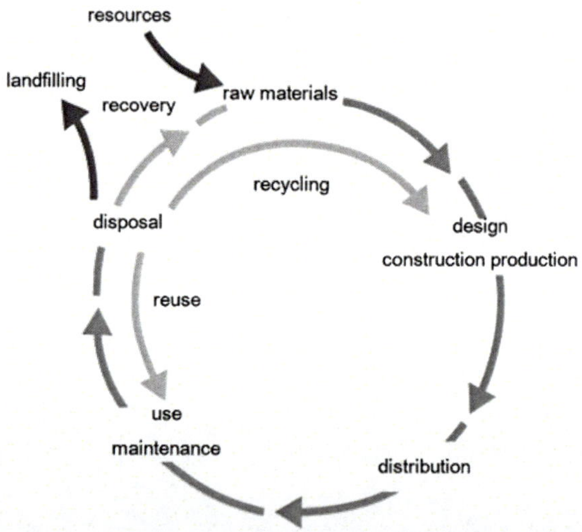

Fig. 5.1 Product life-cycle diagram example

makes us able to rethink and organize the economy as a continuous cycle where nothing can be wasted and the value creation can be maximized eliminating the negative externalities [5, 6]. In this view, any process could be better aligned with circular economy thinking if the initial product and the one from recycled material were both included in a life-cycle assessment perspective. Looking at a product, just as an example, Fig. 5.1 shows a typical life-cycle scheme from extraction and processing of raw materials, design, construction or production, distribution, use, reuse, maintenance and recycling to its final disposal.

The reduction of resource use and air emissions are some of the main goals to be pursued to minimize impacts on the environment, to avoid their transfer from one life-cycle stage to another one, and to achieve performance improvements through the entire useful life [3]. Understanding of the state of the environment by the general community should not be underestimated both for setting of new priorities that can be addressed by design actions, and for translating further environmental needs into consistent designs. Figure 5.2 shows the conceptual framework of the iterative life-cycle design process that includes various sequences of analysis, synthesis and evaluation. Against this background, significant progress has been made in the evaluation of the environmental impacts of road projects through the whole life-cycle [7]; and besides, different life-cycle thinking tools can be now applied to perform an environmental study for infrastructural designs, e.g. concerning highways, but also bridges or individual components and materials used to construct them [8].

In this regard, it is considered appropriate to introduce the Life-Cycle (LC) analysis of an infrastructure such as the assessment of its condition in the remaining life without considering the estimated costs of maintenance, failure, and environmental costs, while the Life-Cycle Cost (hereinafter LCC) analysis also includes estimated

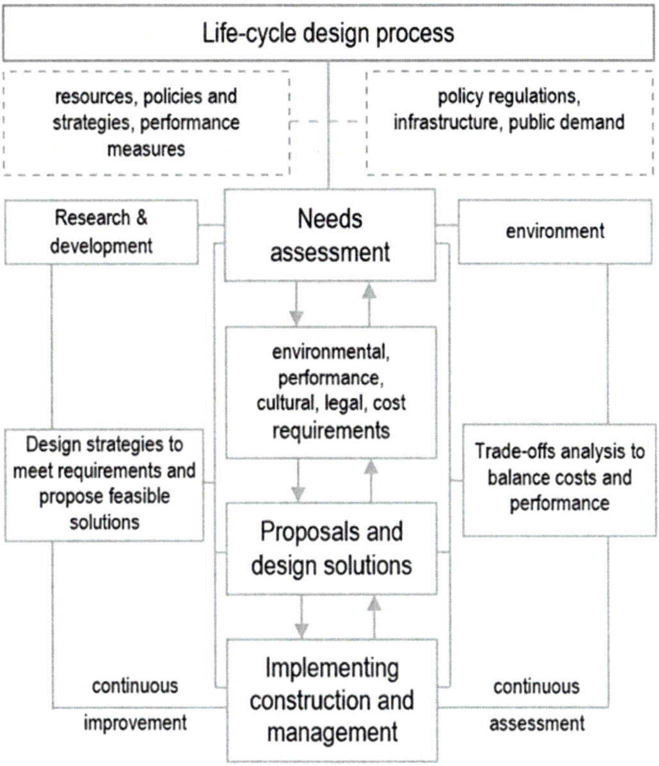

Fig. 5.2 Life-cycle design process

maintenance and failure costs [9]. A common framework for the consistent application of LCC across Europe has been developed for building construction and highways projects without replacing country-specific decision models and approaches; it can be found in [10] where interrelationship between LCC and further sustainability analysis are also given. In this view, the LCC methodology may support the assessment of sustainability of the built environment by providing a systematic approach for comparing all costs combined with constructed assets on a comparable basis since the very early stages in a project [10]; however, its implementation in environmental decision-making is still limited. Although it is not yet possible a fully integration between the results of LCC and environmental analysis, it is important to ensure that the cost implications of environmental impacts are considered and understood in a LCC analysis. Thus, the identification of the best solutions in both economic and environmental terms can be done [11, 12].

Table 5.1 shows an overview of the major applications of LCC methodology, while Table 5.2 summarizes key steps as identified by [10] for common LCC application in EU countries.

Table 5.1 A brief summary of common applications of LCC analysis

LCC uses	Scope	Application stage
Preliminary investigation	Assessing all life-cycle costs to compare alternative design options	– Preparation of strategic cases, – As part of strategic option assessment process, – Pre-planning or early planning stages of a project
Detailed assessment of a project/asset	Assessing the life-cycle costs of an asset for supporting design/investment decision making	– Specific design stages, – Completion or purchase of an asset, – Prior to carrying out remodeling works
Detailed assessment at system or component level	Assessing the life-cycle costs of one or more options	– Detailed design stages, – During operational phase prior to maintenance or disposal stage

Table 5.2 A step by step conceptualization for LCC application across EU countries

Phase of the LCC	Outcome
General identification	Contextualizing of the conditions for implementing actions and understanding of expected achievements
Identification at initial stage	Focusing on the scale of application and related stages, relevant information or specific requirements
Study of the relationships between sustainability analysis and LCC	Understanding of outputs from sustainability assessment that could be inputs into the LCC process and outputs of the LCC process that could be inputs into the sustainability assessment
Choice of the time period and economic evaluation methods	Providing justification on the choice of the period of analysis and investment options assessment methods
Identification of project requirements	Definition of the scope of the project and the key features, project constraints, performance requirements, project budget and timescales, LCC timing into the project plan
Identification of the options and cost items	Understanding of the elements to be subject to LCC analysis, the options for each element to be analyzed and cost items to be included
Assembling cost data	Identification of the relevant costs and their values and time-related data as service life and maintenance data
Calculation of financial parameters	Verifying the sustainability of selected options and calculation of results, application of financial parameters within the cost breakdown structure if used
Economic evaluation	LCC analysis and results
Providing initial and final results	Interpretation of the initial results, identification of needs for further iterations of LCC, final results

When urban space is being analyzed, the definition of an explicit and shared decision-making system can conflict with the unpredictability and uncertainty connected with the evolution of complex contexts [13, 14]. A project (from planning to design level) is not always able, among others, to guarantee the achievement of prearranged goals and to consider all the relations of influence made by its implementation, as well as to avoid the generation of unfavorable externalities [15]. However, the notion of circular cities is already having an effect on the way in which the essential cycles of production and distribution for cities will have to be considered, since cyclic processes characterize the city development's nature and resource allocation between the urban activities; see e.g. [16–21].

5.1.1 The Background of Circular Cities

Transition towards the circular paradigm needs that all infrastructures on which urban subsystem operates should be (re)designed and cities should be also planned so that land and infrastructures can be re-used over time [10, 22]. If it is true that the future cities will have to be circular to be sustainable cities, it is also true that circularity of products or services should be properly quantified to better plan urban strategies and to support decision making [23]. This is important when the impacts (or benefits) of alternative design proposals should be evaluated and intervention priorities on the base of the relevant impacts should be established.

Life-cycle methods have been applied to various issues involved in urban metabolism, but there is still a need to study the short-comings of their application to aspects related to sustainability of cities. The environmental issues in the design of a facility whose life-cycle also includes a disposal phase, have gradually requested to balance functional and disposal requirements right from the facility's planning and design phase. In this regard, the Life Cycle Assessment (LCA) in its traditional conception considers the entire life-cycle of the product or service from raw material sourcing to manufacturing, distribution and use [24]. Despite criticisms against it, this method can distinguish good circular economy ideas from bad ones and can be a decision support tool to provide an effective contribution towards greater sustainability of a product or service [25]. Abraham and Dickinson [26] developed a model to consider the facility's changing conditions, to track the changes in the disposal costs, to qualify and quantify the facility's final disposal cost; however, major conclusions only concerned the role of LCC analysis in the feasibility study of construction projects, while more case studies should be provided, particularly in the public sector; see also [27]. However, except for bridge management [28], design of structural components [29, 30], highway maintenance and rehabilitation strategies [31], there is still poor application of life-cycle costing methods for the construction projects especially at urban level. Fuller and Petersen [32] were among the first to distinguish the Life-Cycle Cost from Life-Cycle Cost Analysis (LCCA) that is the comparative assessment of more design alternatives over an agreed period of analysis, identifying whether the analysis is for only a part or for the entire life-cycle of the constructed asset over its assumed life. They concluded that the LCCA usually rules out environmental costs, while the similar LCC or LCA can usually include them. Focusing on local road infrastructures, Goh and Yanh [33] developed a life-cycle costing analysis model for sustainability enhancement in road infrastructure projects. Despite its potential the proposed model used sustainability indicators that were locally identified; thus, in order to reach more generalizable conclusions, the authors stressed the fact that further study had to be done. Brilon [34] also expressed his point of view and argued that a 'whole year analysis' on the facility's use for assessing the economic benefits from investments on road capacity could give a more complete picture of the consequences caused by different design options to the decision makers.

In order to prompt policy shifts towards a system perspective, further LCC applications have been about road planning and management [12, 35], transportation investment decisions [36], pavements [37]. A specific a life-cycle costs estimation tool has been recently developed to address comparison of alternative designs for existing and new intersections [38]; however, its application is still limited to a series of US test cases. At last, the Social Life Cycle Assessment completes the pillars of sustainability by considering the social aspects. However, Social LCA (S-LCA) and life cycle sustainability assessment (LSCA) are newer methods and their application is still incipient and needs standardization [39, 40].

5.1.2 The Aim of the Study

In the present study a project example has been developed for comparing three alternative intersection designs based on their total life-cycle costs. Consistent with [38], the costs to be included in an analysis directed toward deciding among alternative investments, included agency costs, users and non-user costs. A roundabout and a signalized intersection have been conceptualized and then compared to an existing Two-Way-Stop-Controlled intersection located in Palermo, Italy; it represented the base case.

In order to identify the preferred configuration, the Life-Cycle Cost Estimating Tool proposed by [38] has been applied to compare the alternative intersection projects and to estimate the total net present values of benefits and costs over the life-cycle of each of them. The study allowed us to identify the most appropriate data sources to be used and performance measures to be included in the life-cycle cost approach. The results confirmed the feasibility and efficacy of the methodology being used to make a relative comparison among the intersection alternatives. Consistently with the stated objectives, this study aims to contribute to show the functionality and value of the spreadsheet-based tool proposed by [38] to potential users and gives a contribution to the LCCET validation process. Although this study makes no claim to be exhaustive, the results express an Italian point of view to facilitate its implementation in territorial contexts different from those of validation.

5.2 The Life-Cycle Costing Method for Evaluating Urban Intersection Designs

Based on real-world experience, different useful life or differences in the useful life of components as pavements, bridges and other structures or street furniture equipment can characterize the infrastructure construction projects as highways, roads, streets, interchanges and intersections. It is further noted that each component could be also efficient beyond the useful life or the analysis period under examination. There could be an instance where alternative designs are characterized by varying spatial scopes with respect to the interaction among the parts that comprise each project, or new intersection types have not been considered but should have been considered among the alternatives that were evaluated. In this view, an analysis method is needed to collect the available information, to determine the performance measures and then the life-cycle costs of each project so as to facilitate the comparison among the design alternatives. The Life-Cycle Cost analysis of road facility and intersection designs include not only the construction costs and other costs incurred prior to and during construction, but also the costs of operating, maintaining and preserving the project in the immediate and long term, and at higher efficiency levels, throughout its life-cycle. This also requires that there are the financial conditions to ensure that the infrastructure can be replaced or renewed at the end of its useful life also in response to increases in traffic demand. In this view, estimating the life-cycle costs (and benefits) of alternative designs is the necessary precondition for really understanding that costs can be assessed and properly compared when they are related to particular levels of service [10, 38].

Estimation of the life-cycle cost and benefits of intersection designs means having the present values of project's costs and benefits expressed in monetary value. Thus, a discount rate should be applied to convert future costs (and benefits) into current values and to place them on a par with costs (and benefits) incurred today for making direct comparison. Incorporating LCC into decision-making methodologies will make easier to assess and to compare intersection designs with different life spans, or varying spatial scopes, and to differentiate the salvage and terminal value of each design. More in general, it will demonstrate that the best value for money across the asset life-cycle can only be assured by purchasing green and socially preferable alternatives. A cost-based analysis should be also scalable over different life cycles for different designs.

When various design alternatives with varying useful lifespans should be compared, identification of each lifespan is useful for defining the time period over which the analysis will be made. Many factors may influence the choice of the analysis period as the growth context with regard to effects on traffic and capacity of the intersection, the expected useful life of its components or an agency planning horizon. In any case, a long period of analysis may not always add useful information from an economic point of view, even for uncertainties about traffic growth, technologies and their costs; in turn, a short period of analysis could make difficult to capture the effects of important life-cycle differences of the design options and

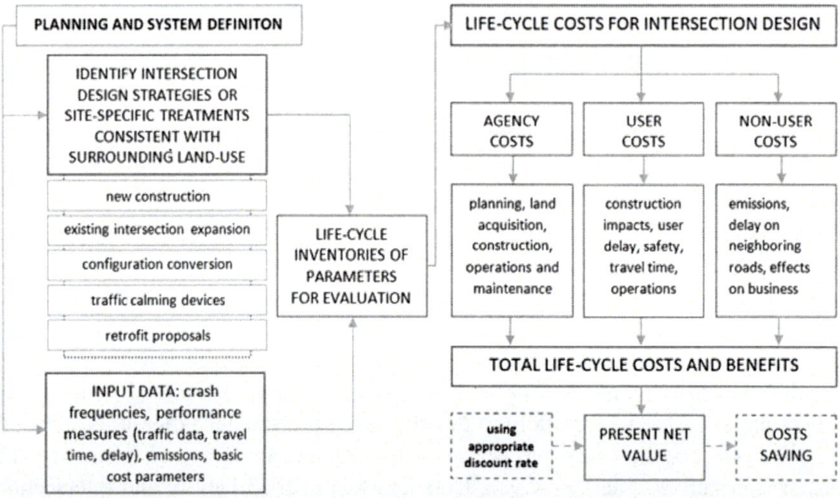

Fig. 5.3 An overview of the life-cycle cost-benefit approach for evaluating intersection designs

to estimate terminal or salvage values to eventually incorporate benefits beyond the lifespan of each design.

As introduced above, a Life-Cycle Cost Estimation Tool has been recently developed to address comparison of alternative designs for existing and new intersections thorough a common method of analysis [38]. Figure 5.3 shows an overview of the life-cycle costs model for evaluating the sustainability of different design solutions for intersections. The abovementioned tool can use input data derived from other analysis tools such as Highway Capacity Manual [41] or traffic micro-simulation for performance measures, and Highway Safety Manual [42] for crash forecasting. Further inputs include the discount rate that has to be used to convert all costs into present values and unit costs for the performance measures [43, 44]. Since its application is still limited to a series of US test cases, the case study presented in this chapter can represent an Italian point of view which contributes to fill the existing gap on the topic.

Note that the costs to be included in an analysis directed toward deciding among alternative investments, include agency costs, users and non-user costs; as well as benefits, they could be calculated in each year through the analysis period especially where high growth of traffic demand is expected. The Agency costs include planning, permitting, regulatory review costs, that become sunk costs (if already incurred), engineering costs related to the project once it has been defined, the cost of acquiring the land and access, construction, operations and maintenance costs incurred throughout a project's life-cycle.

In turn, user costs include costs due to impacts during construction, delay experienced by users through each intersection alternative, travel time reliability since travelers have to budget additional time to reach their destinations at the desired time; further, safety costs should be included to consider the number and severity

Table 5.3 Summary of LCC applications to road design and interventions

Area of analysis	Study or application	Territorial body		
		State	Region	City
Planning policies, programming & analysis, budget analysis	– Funding functional improvements for intersections within a general redevelopment program	✓	✓	
	– Funding specific intersection treatments	✓	✓	
Area-wide, road network or corridor	– Analysis at network or corridor level	✓	✓	✓
	– Ranking of alternative intersection design or treatments	✓	✓	✓
	– Signal timing or retiming	✓	✓	✓
Single intersection	– Upgrade or conversion	✓	✓	✓
Maintenance	– Pavement markings, replacing of damaged, or missing signs, replacing of barriers, roadside safety devices and drainage features	✓	✓	✓
	– Upgrading the types of costs or performance metrics	✓	✓	✓

of crashes for a given intersection design, as well as vehicle operating costs. The conversion from one layout to another one can induce effects in terms of changing the traffic demand or the route traveled by users not only through the intersection under examination, but also elsewhere on the road network. As a consequence, non-user costs can arise and include delay to travelers due to change in traffic demand or travel times, further emissions from vehicles with effects on health and climate change; see Table 5.3 for some possible applications of LLC by area of interest.

5.2.1 *Materials and Data*

The case study got started with the identification of the configuration of the two-way stop-controlled intersection between Levriere Road and Ermellino Street located in Bonagia neighborhood in Palermo, Italy. Since there was no evidence of operational interaction with the adjacent intersections, the existing intersection (the base case) was treated as an isolated road system. Two intersection alternatives were first conceptualized and then compared to the base case: a roundabout (alternative 1) and a signalized intersection (alternative 2). In order to identify the preferred configuration among the proposed design alternatives, the Life Cycle Cost Evaluation Tool (LCCET) has been applied to estimate the total net present values of costs over the life-cycle of the alternative treatments [38]. Note that this tool offers great flexibility to the user because he has more options to estimate some of the performance measures that are used as inputs, to define his own cost categories and to override any of the default costs or other parameters, and so on. In order to provide a common base of comparison, cost data given in local currency were converted in dollars averaging the exchange rates of the last five years. The reader is referred to [38] for more information about how the tool works.

With a view to providing input data and then cost information for the base case and the alternatives 1 to 2, the methodology consisted of eight steps as follows:

Step 1: choice of the analysis period for the study,
Step 2: evaluation of the initial and the final traffic scenarios,
Step 3: conceptualization of the intersection design alternatives,
Step 4: delay estimation under traffic demand,
Step 5: safety performance measurement from crash data,
Step 6: preparation of the road-maintenance program and worksite organization,
Step 7: estimation of the emissions from vehicle operations,
Step 8: calculation of the cost items and Net Present Value for the design alternatives.

Step 1 consisted of choosing the analysis period to have a common comparison base among the design alternatives under consideration. A reference period of 25 years from the year 2019 until 2044 has been chosen. This choice was consistent with what proposed by [38] with regard to growth conditions in the environmental context where the existing intersection is installed and to the useful lives of components of each alternative design; it was also made in analogy to periods of analysis used for similar road infrastructure projects; see also [45].

Step 2 consisted of evaluating the initial and final traffic scenarios. Total entering volumes were collected in the field on weekdays from Tuesday to Thursday in April and May, 2019; they were videotaped during daily peak and off-peak hours in 15 min intervals (i.e. morning peak hour 7:30 to 8:30 a.m., afternoon peak hour 12:30 to 1:30 p.m., evening peak hour 6:30 to 7:30 p.m., morning off-peak hour 10:00 to 11:00 a.m., afternoon off-peak hour 3:00 to 4:00 p.m.) at the existing TWSC intersection. By way of example, Fig. 5.4 shows the traffic load diagram relating to the number of vehicles entering the intersection within the evening peak hour (6:30 to 7:30 p.m.); it was the most critical condition where a traffic volume of 1680 veh/hr entered the intersection with more than 80 percent made by passenger cars.

Note that the demand parameters worksheet provides traffic demand profiles by facility type and converts average annual daily traffic into hourly demands so as to give inputs in a consistent format; the user is able to override or modify default data to reflect real local conditions or to incorporate traffic data collected in the field. The same AADT values and annual truck percentages in the initial year and the end year were used for each intersection alternative. In order to estimate the amount of traffic in the interim years and the final year, the traffic data in the Municipality of Palermo from 2012 to 2018 were linearly interpolated and used to determine the traffic growth rates by type of vehicle and the total fleet [46, 47]. The demand parameters were the

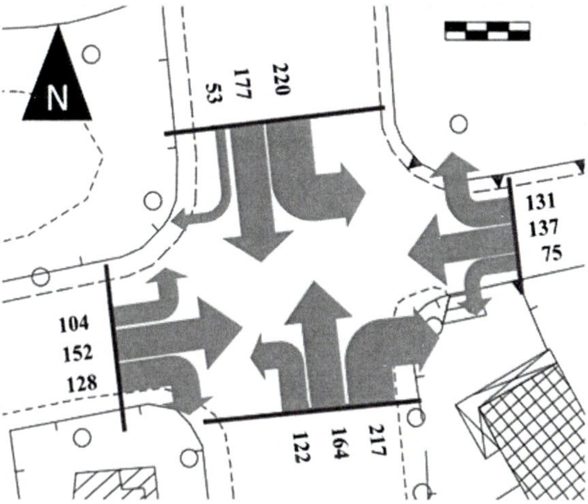

Fig. 5.4 Traffic load diagram for the base case in the evening peak hour

AADT of 18,667 [veh/d] for the initial year and of 20,834 [veh/d] for the end year, while the percentages of annual heavy vehicles were among 6.4% and 8.0% for the initial year and the end year, respectively. No assumption was made for volumes of bicycle and pedestrians in view of their small amount.

Step 3 consisted of defining the design alternatives to be proposed and evaluated. Some assumptions were made in order to conceptualize a base case with which the intersection alternatives could be compared. Thus, the following intersection design alternatives were considered:

- base case (alternative 0): there was conservation of the original layout of the existing TWSC intersection without any interventions on geometric design or traffic control; the project started only consisted on a road pavement renewal,
- roundabout (alternative 1): consistently with [48], the project consisted on the construction of a compact single-lane roundabout with single entry and exit lanes. The project maintained all turning movements and reduced the number of conflict points by changing crossing conflicts into merging conflicts; it also moderated speeds on entries and through the circulatory roadway by providing enough deflection. The reduction in speeds of vehicles and the removal of head-on and angle crashes also reduced the crash severity. The project also consisted on a modernization of the street section, realizing a flexible pavement,
- signalized intersection (alternative 2): the project consisted on the conversion of the existing intersection into a signalized intersection with a dedicated left turn lane and a shared through-right turn lane on the northbound and southbound entries, and with a single lane on the eastbound and westbound entries. The project included a 2-phases, fixed-cycle traffic signal with a cycle time of 61 s.

Figure 5.5 shows the conceptual layouts and geometric details of the intersection alternatives being evaluated. Table 5.4 summarizes input data for the cycle

(a). Base case (Alternative 0)

Entry	Lane*	Lane width [m]	Parking	Grade [%]
East	1	4.75	No	0
West	1	4.00	Both sides+	0
South	1	5.00	Both sides ++	0
North	1	6.50	no	3

(*) One-way entry (and exit) lane; (+) Longitudinal type; (++) 26 m from the entry line.

(b). Roundabout (Alternative 1)

Geometry	[m]
Outer diameter	26.00
Circulatory roadway width	7.00
Splitter island width	5.00
Entry lane width	3.50
Exit lane width	4.50
Deviation angle between subsequent entries	> 45°

(c). Signalized intersection (Alternative 2)

Traffic signal phasing	[s]
Cycle time	61
Amber time	4
All red time	3
Effective green (I phase)	17
Effective green (II phase)	30

Note: one-lane entry (and exit) 3.50 m wide at eastbound and westbound entries; two-lane entry each 3.00 m wide at southbound and northbound entries, while exit lane was 4.00 m wide.

Fig. 5.5 Conceptual layouts of the intersection design alternatives

Table 5.4 Data summary for the cycle phasing

	Lane	Movement	15 min peak flow rate	Saturation flow [veh/hr]
Phase 1 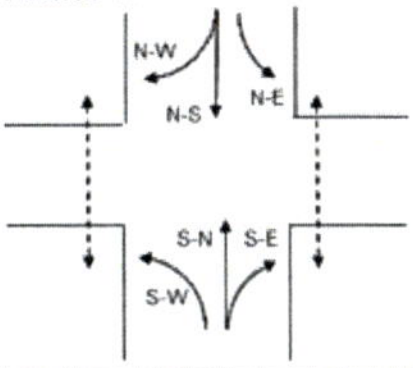	1	W-N	109	1,461
		W-E	160	
		W-S	135	
	2	S-W	128	1,741
		S-N	173	
	3	S-E	228	1,802
Phase 2 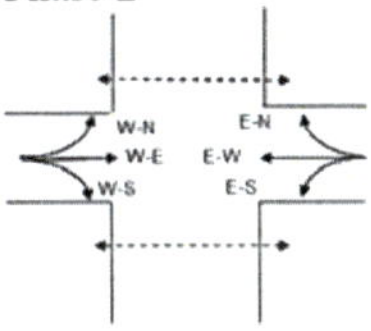	4	E-S	79	1,701
		E-W	138	
		E-N	144	
	5	N-E	232	1,809
	6	N-S	186	1,809
		N-W	56	

phasing for the alternative 2 and the two-phase design that also incorporated pedestrian protection; calculations were based on procedure reported by [41]. Beyond this, the comparison among the design alternatives included the modernization of the street section by realizing a flexible pavement with service life and characteristics of layers different for each design option. Table 5.5 shows the characteristics of pavements for each design alternative based on what experienced for similar Italian roads at design-level or in operation under traffic. The lifespan of the flexible pavement of each design alternatives in Fig. 5.5 was defined based on the functional adequacy of each of them within a time period beyond which the road pavement could become obsolete to serve the traffic demand; however, land-use patterns and changes in traffic volumes and operational performances could make pavements obsolete also before the end of the structural life.

Table 5.5 Summary of characteristics for flexible pavements

Alternative		0	1	2
Lifespan (years)		15	20	10
Layers of material [mm]	Surface course	40	50	40
	Binder course	50	60	50
	Base course	90	80	80
	Sub-base course	330	350	300
Design ESALs[1]		1,435,671	2,307,039	980,891

[(1)]ESAL stands for the equivalent single axle load that is a mixed traffic stream of various axle configurations and axle loads predicted over the design period, and converted into the standard 80 kN single axle loads summed over the design period [49–52]

Table 5.6 Control delay and level-of service determination for the design alternatives

Entry	Movement	Pm control delay [s] (level-of-service)					
		Case base		Roundabout (Alternative 1)		Traffic signal (Alternative 2)	
		2019	2044	2019	2044	2019	2044
West	West-North	8.4 (A)	8.7 (A)	12.1 (B)	16.3 (C)	18.4 (C)	22.6 (C)
	West-East	0.0 (A)	0.0 (A)	12.1 (B)	16.3 (C)	18.4 (C)	22.6 (C)
	West-South	0.0 (A)	0.0 (A)	12.1 (B)	16.3 (C)	18.4 (C)	22.6 (C)
East	East-South	8.2 (A)	8.5 (A)	16.0 (C)	24.0 (C)	17.2 (C)	21.6 (C)
	East-West-	0.0 (A)	0.0 (A)	16.0 (C)	24.0 (C)	17.2 (C)	21.6 (C)
	East- North	0.0 (A)	0.0 (A)	16.0 (C)	24.0 (C)	17.2 (C)	21.6 (C)
South	South-West	25.2 (D)	29.6 (D)	25.1 (D)	31.1 (D)	22.3 (C)	24.1 (C)
	South-North	30.9 (D)	35.0 (D)	25.0 (C)	31.1 (D)	35.9 (E)	40.4 (E)
	South-East	43.5 (E)	>E	25.1 (D)	31.1 (D)	35.9 (E)	40.4 (E)
North	North -East	>E	>E	13.9 (B)	19.1 (C)	24.0 (C)	29.2 (D)
	North-South	40.9 (E)	>E	13.9 (B)	19.1 (C)	35.0 (D)	39.9 (E)
	North-West	21.7 (C)	22.8 (C)	13.9 (B)	19.1 (C)	35.3 (E)	39.9 (E)
Intersection		35.7 (E)	46.0 (E)	17.3 (C)	26.1 (D)	20.3 (C)	29.4 (D)

Step 4 consisted of estimating the intersection control delay for the initial year (2019) and the end year (2044) for the alternative design options; again, calculations were based on the HCM procedure [41] depending on the delays associated with the individual maneuvers.

By way of example, Table 5.6 shows the main results concerning pm peak hour control delay and the level-of-service determination.

A square-shaped study area with side of 235 m was identified so as changes in traffic characteristics outside its cordon line had no significant effect on traffic operations within it. The positioning of the boundaries was done by analyzing intersections of analogous size where drivers covered similar travel distances [47]. The study area was sized by overlapping the footprint of each intersection alternative; the cordon boundary was also set to account for the deceleration distance of vehicles approaching each anticipated queue.

Table 5.7 shows information summary of queue lengths calculated using the HCM procedure [41], while Fig. 5.6 shows delay data with reference to the morning peak hour and afternoon peak hour for the initial year and the end year of the analysis period; they were then used as input for the LCCE tool.

Step 5 consisted of estimating the expected crash frequency for each intersection in the reference period. Since user safety had to be monetized, the present value for each crash category was totaled annually for each year of the time horizon of the project. Estimation of the expected crash frequency per year allowed us to determine the annual cost of crashes for each design alternative and then to assess, in monetary

Table 5.7 Information summary of queue lengths for the design alternatives

Entry	Movement	Queue lengths [m]					
		Case base		Roundabout (Alternative 1)		Traffic signal (Alternative 2)	
		2019	2044	2019	2044	2019	2044
West	W–N, W–E, W–S	0.0	0.0	11.0	17.0	13.0	15.0
East	E–S, E–W, E–N	0.0	0.0	8.0	10.0	11.0	12.0
South	S–W	19.0	31.0	22.0	39.0	21.0	29.0
	S–N, S–E	19.0	31.0	22.0	39.0	8.0	9.0
North	N–W, N–S	26.0	56.0	11.0	16.0	16.0	14.0
	N–E	26.0	56.0	11.0	16.0	12.0	16.0

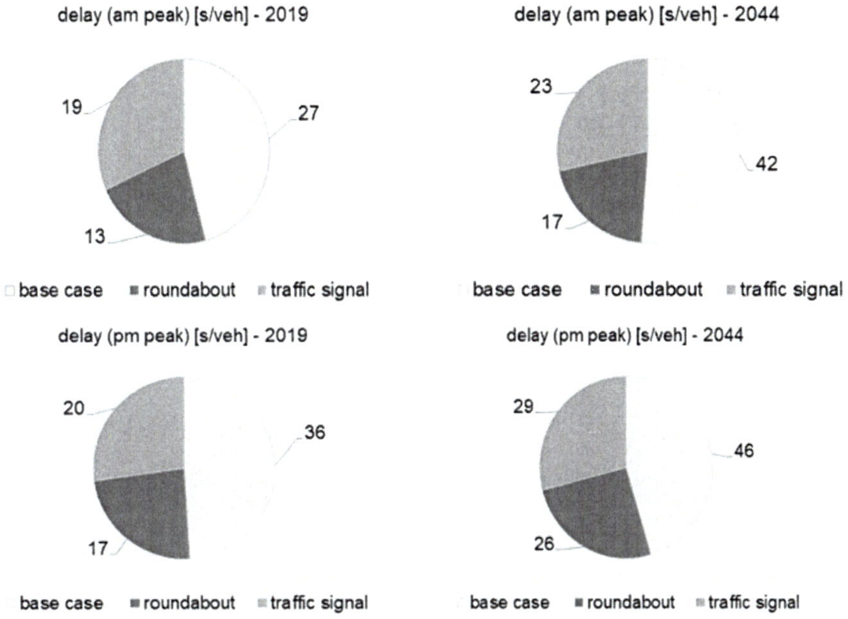

Fig. 5.6 Delay by design alternative for the initial year and the end year

terms, the resolving effectiveness of each design alternative compared to the base case from a road safety point of view.

Crash history was obtained from the local Police database for the years 2012 to 2018. In this step the predictive method proposed by [42] was applied to estimate the expected crash frequencies; the safety performance function proposed by [53] was used for the roundabout alternative. Note that only total and injury crashes for each alternative were considered since no fatal crashes occurred within the time period on which the analysis was based. This is consistent with the crash history in Palermo

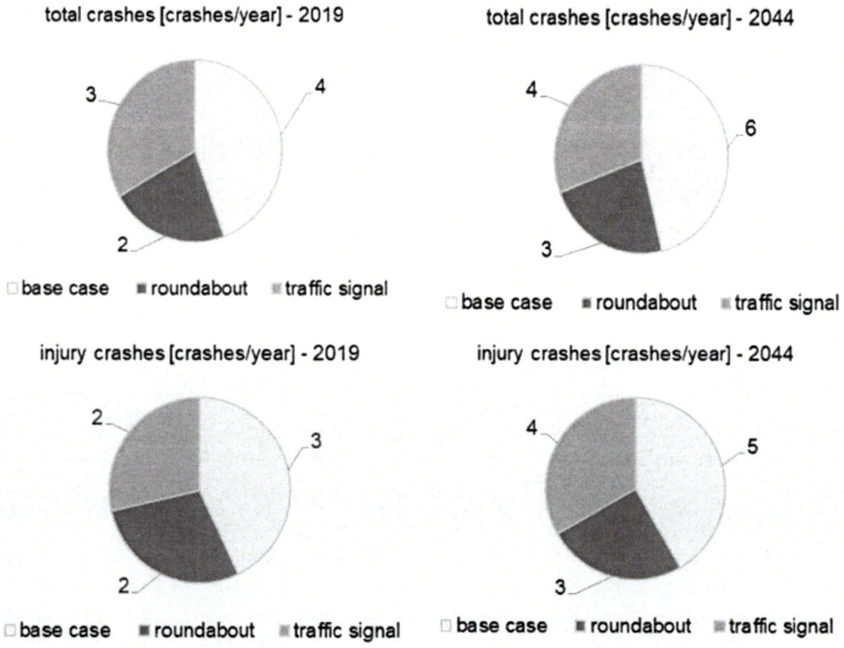

Fig. 5.7 Total and injury crashes by design alternative for the initial and end year

City where a frequency of 0.4 fatal crashes/year was observed in 2012–2018 years. Figure 5.7 shows total and injury crashes by design alternative for the initial year and the end year of the analysis period.

Step 6 consisted of developing the road—maintenance strategies to preserve each design alternative overtime. Given that construction and maintenance costs are incurred throughout the infrastructure's life cycle, the definition of the number of years to complete maintenance works and pavement maintenance strategies were based on the decay curves of the friction coefficient and pavement service life [49, 51]. In order to be consistent with the objectives of the chapter as introduced above, details on construction machinery considered for the planning of road construction works and the enterprise configuration in the various projects were not introduced here; for further details see again [47].

Cost information for the planned works has been derived from the official single price list of Sicily, Italy [52], where the unit costs of different productions envisaged for the various activities can be found. The costs of supply of raw materials and landfilling activities of waste materials have been determined based on the distance between the construction site and potential suppliers of building and road materials, as well as firms performing services for removal of special waste closer to it. Table 5.8 summarizes the maintenance strategies concerning the planned works for each design alternative and the costs mentioned in U.S. dollars at 2019.

Table 5.8 Summary of road—maintenance strategies and costs for the design alternatives

Design alternative	Intervention	Year	D* [days]	Costs [$]
Base case (alternative 0)	Pavement renewal	2019	19	361,139.37
	Double surface dressing	2026	10	207,817.11
	Pavement renewal	2034	19	361,139.37
	Resurfacing	2042	12	291,791.39
			Total	1,221,887.24
Roundabout (alternative 1)	Pavement construction	2019	30	449,965.79
	Slurry seal	2026	4	35,145.60
	Resurfacing	2034	12	329,418.49
	Pavement renewal	2042	19	382,833.31
			Total	1,197,363.19
Signalized intersection (alternative 2)	Pavement construction and traffic light system	2019	25	382,469.59
	Single surface dressing	2025	5	111,462.44
	Pavement renewal	2030	19	345,047.86
	Micro surfacing	2036	9	132,250.99
	Pavement renewal	2042	19	345,047.86
			total	1,316,278.74

Note using a work breakdown structure, D is the sum of the duration of all the construction and maintenance works: $D_j == \frac{Q_{tot,j}}{Q_{h,j}}$, where is the total production related to the activity j [hr] and $Q_{h,j}$ is the hourly production of the instrumental factors of the work j

Once the maintenance strategies of the projects and the costs of the planned interventions were known, the salvage value at the end of the pavement lifetime for each intersection alternative has been determined as follows:

$$SV_f \cong SL_j = C_f(t_n) * \frac{e_n - (t_f - t_n)}{e_n} \tag{5.1}$$

where:

SV_f is the salvage value of the alternative j in dollars,

SL_j is of the remaining useful life for the pavement of the alternative j in dollars,

t_f is the end year of the reference period (*years*),

t_n is the year that the last extraordinary maintenance work is made (*years*),

e_n is the resolving efficacy of the last extraordinary maintenance work (*years*).

The salvage values were determined and expressed in terms of the monetary value of the year 2019 where the economic analysis was made [47].

For what concerns the worksite organization an entry at a time was considered, shutting down half the road platform and putting an alternating one-way system in the other half of it. The use of a mobile three-phase traffic light was also planned in a working area that is about 110 m long at each entry; a speed limit of 30 km/h on the alternating one-way road segment was imposed. A further condition was that the last vehicle before the red light had to have the time to run the entire alternating one-way road section and to cross the intersection before the green light shot for vehicles coming from the other entries without worksite. A cycle length of 105 s resulted appropriate to minimize delays and queue lengths.

Step 7 consisted in estimating the emissions from vehicles to monetize their costs. And while it is true that several pollutants from vehicles have to be considered because of their health effects at local level, it is also true that greenhouse gases should be considered and monetized because of their effects on climate change. This aligns with the European long-term strategy of transition to green energy technologies and decarbonization so that the whole transport system becomes climate- and environmentally-friendly, resilient, resource-efficient and safe for the benefit of all citizens and the general economy [54]. Emissions data were entered in the Life-Cycle Cost Estimation Tool as annual CO_2 equivalent emissions in metric tons and annual tons of emissions per year by type of criteria pollutant; emissions data were entered for the same years as those for delay data. The costs were then obtained by multiplying the estimated quantities by the unit costs of GHG emissions and other pollutants (e.g. particulate matter and nitrogen oxides); information on costs may be found in [55, 56] as referred by [38].

In order to estimate the daily emissions and then the corresponding costs, traffic through the intersection was simulated during the whole day by using the Italian microscopic traffic simulation model Tritone (version 2016) [57]. The software above presented in that package version the model developed by [58]. Based on the geometry in Fig. 5.5 the network model of each intersection was built; the urban speed limit of 50 km/h was set as the maximum speed at which each element of the network model can be traveled under free flow conditions. The study area was chosen as introduced above. Within the simulated vehicle fleet there were 45% of diesel-powered cars, 45% of petrol-powered cars, 9% of LPG-powered cars, 0.5% of hybrid cars and 0.5% of electric cars for the initial analysis year (2019). Two vehicle fleets were considered for the end year (2044): (a) the same vehicle fleet composition in the 2019; (b) the vehicle fleet composition with 25% of diesel-powered cars, 30% of petrol-powered cars, 20% of LPG-powered cars, 15% of hybrid cars and 10% of electric cars. The reason for the dual choice was made to consider the progressive transition to electric vehicles, and to isolate and to test their effects on emissions; it is well-known that electric vehicles emit almost three times less, on average, CO_2 than petrol or diesel cars, and might reduce CO_2 emissions four-fold by 2030 thanks to an EU grid relying on renewables [59–61]. Table 5.9 shows the vehicle type attributes of the simulated fleets. The GEH index resulted smaller than 5 in more than 85% of the cases when the simulated flow rates were compared with the on-field detected flow rates at each entry throughout each 15-min sampling interval in the pm peak hour [62].

Daily emissions were estimated based on the hourly distribution of the average daily traffic; the hourly traffic profile of the functional class of urban minor arterials proposed by [38] was applied. The simulation outputs were the mean value of the

Table 5.9 Summary of the characteristics of the fleet

Vehicle type	Vehicle characteristics						
	Power type	Length [m]	Width [m]	Max speed [km/h]	Max acceleration [m/s^2]	Max deceleration [m/s^2]	CO_2 [g/km]
Car[1]	Petrol	3.90 (2.70; 4.84)	1.74 (1.49; 1.94)	173 (130; 250)	2.28 (1.20;4.79)	3.41 (2.40; 4.00)	0.14 (0.088; 0.36)
	Diesel	4.23 (3.55; 4.83)	1.77 (1.63; 1.91)	191 (164; 230)	2.58 (1.84;4.27)	5.24 (3.68; 8.55)	0.17 (0.081; 0.98)
	Hybrid	4.34 (3.75; 4.64)	1.81 (1.70; 1.90)	205 (150; 280)	3.14 (1.70; 5.67)	6.74 (3.00; 11.30)	0.13(0.039; 0.288)
	Electric	4.28 (3.80; 4.80)	1.78 (1.73; 1.83)	185 (150; 220)	2.23 (2.16; 2.35)	4.60 (4.32; 4.80)	–
Bus	Diesel	12.00	2.40	100	1.20	7.50	0.50
Heavy vehicle < 3.5 t	Diesel	10.00	2.20	120	2.50	5.50	0.50
Heavy vehicle > 3.5 t	Diesel	10.0	2.60	100	2.00	5.00	0.50
Motorcycle	Petrol	1.45	0.50	220	3.50	7.50	0.50

Note Mean value of the distribution of the parameter with the corresponding minimum and maximum values in round brackets, if available; (1) based on [63] petrol-powered cars were: 5% of EURO 2; 39% of EURO 4; 30% of EURO 5; 26% of EURO 6; (2)EURO standards for diesel-powered cars were: 43% of EURO 4; 36% of EURO 5; 21% of EURO 6; (3) EURO standards for hybrid cars was 100% of EURO 6

concentration (grams) of carbon dioxide (CO_2), nitrogen oxides (NO_x) and particulate matter (PM_x) returned in 4 runs, each of them performed within a control time interval of 1800 s by which the results of the 24-hour micro-simulation were aggregated. Emission estimates by design alternative for the initial year and the end year are shown in Fig. 5.8; consistently with the benefit-cost analysis approach on which the tool is based, they were used as data input to estimate the corresponding costs.

Step 8 consisted in calculating the various cost items over the lifetime of each design alternative and the Net Present Value (NPV) for each of them, so as to be able to compare the design solutions and to identify what best meets the overall needs of the site. The Life Cycle Cost Estimating Tool as introduced above was used to compare the different design alternatives being examined. Since the tool uses the benefit-cost analysis approach, it allowed to estimate the net present values of benefits and costs of each intersection design alternative. The data that are necessary for proper working of the tool are the cost parameters to be used in the calculation, total traffic

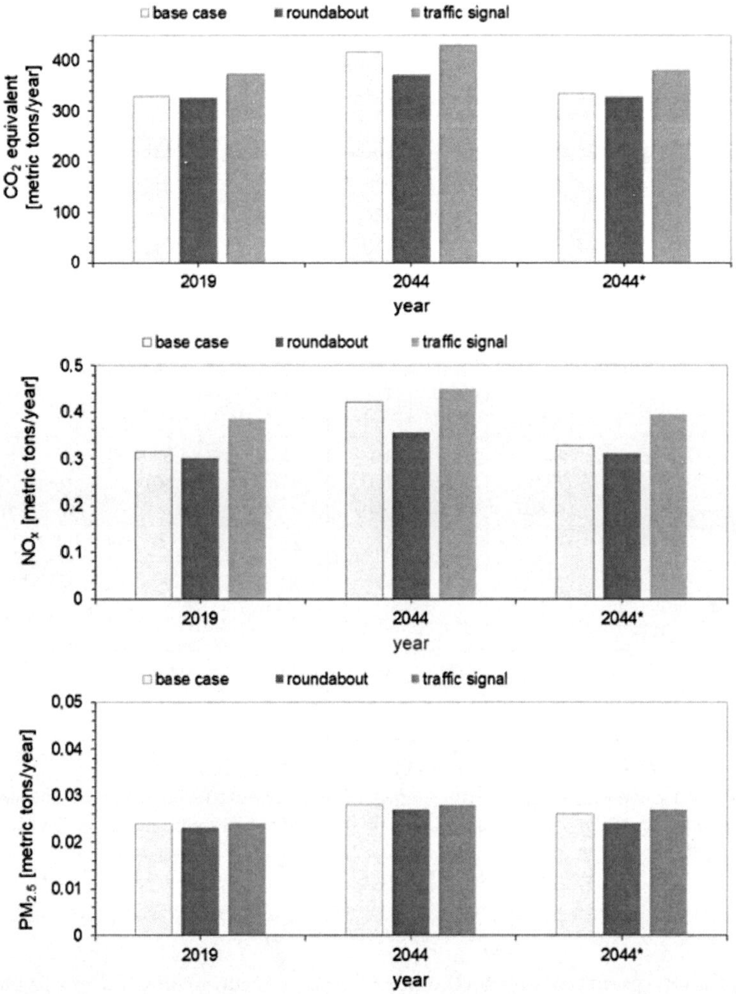

Fig. 5.8 Emission estimates by design alternative for the initial and end year. *Note* Two vehicle fleets were considered for the end year (2044): **a** the same vehicle fleet composition in the 2019; **b** the vehicle fleet composition with 25% of diesel-powered cars, 30% of petrol-powered cars, 20% of LPG-powered cars, 15% of hybrid cars and 10% of electric cars

demand entering the study area in the initial year (2019) and the end year (2044) (i.e. morning and afternoon peak hour volumes in vehicles per hour and annual average daily traffic through the intersection in vehicles per day), average delay in seconds per vehicle during morning and evening peak hours for the initial year (2019) and the end year (2044), total and injury crashes for the initial year (2019) and the end year (2044), CO_2, NO_x and PM_x emissions in metric tonnes for the initial year (2019) and the end year (2044), as well as planning and construction costs. In order to consider construction impacts, calculation of costs also took into account the presence of work zone. Since during construction, users could suffer changes in delay, or incur further operating expenses and so on, the same monetization process needs insight into the type of construction associated with each intersection project. In essence, users bear costs of work zone-related delay and vehicle operating costs generated by changes in driving regimes that have to be made. In order to calculate increases in delays and queue lengths, traffic conditions with and without work zone have been simulated for each intersection alternative and each year in which interventions were planned; see [47].

In order to assess the feasibility of converting an unsignalized intersection to a roundabout or a signalized intersection, the following cost items were considered: the cost of the increase in the total delay, that is the monetary value of the additional time needed to cross the intersection during work zone operations; the additional vehicle operating cost for no load operation during idling while queued. Information on unit user costs can be also found in [64]. Consistently with [38], financial costs were calculated and discounted to the initial year of the analysis period that is the earliest year in which planning and construction begin for the design alternatives. Figure 5.9 illustrates the costs information summary for every intersection project; the years in which the above costs were incurred have been taken into account.

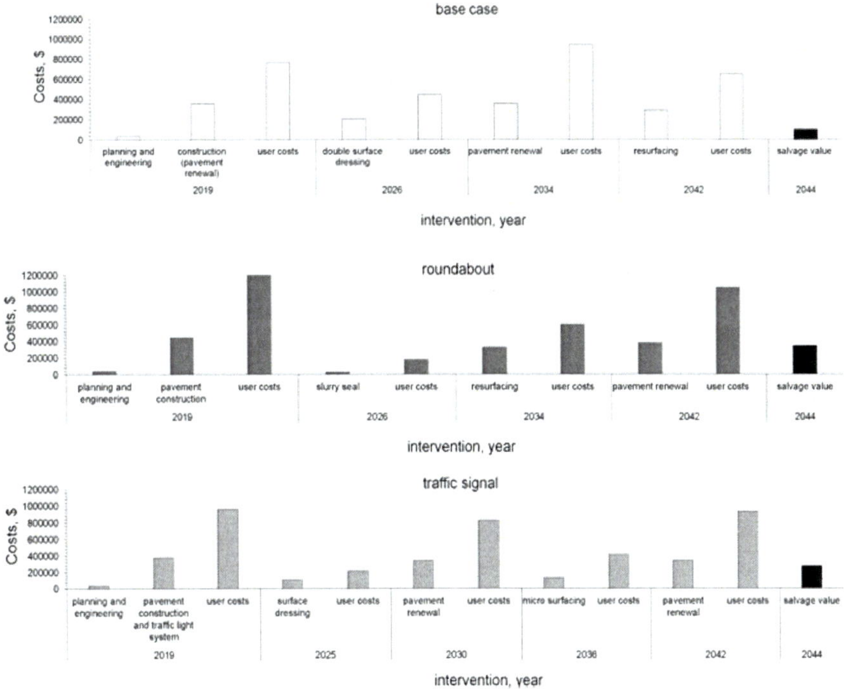

Fig. 5.9 Life-cycle costs information summary for intersection design alternative

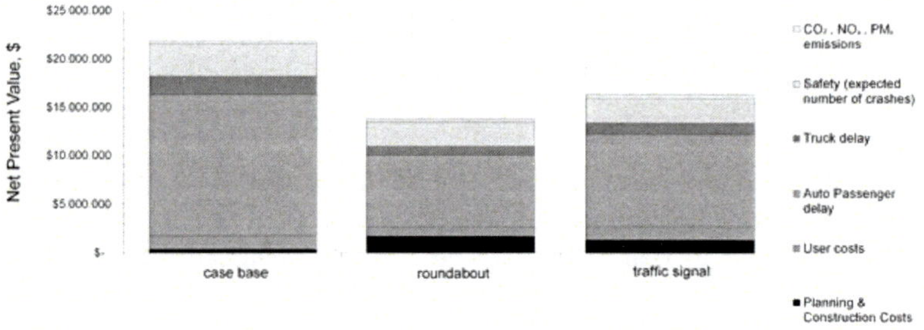

Fig. 5.10 Net present value of total costs from the analysis of the intersection alternatives. *Note* To test emissions, the traffic mix is made by 45% of diesel-powered cars, 45% of petrol-powered cars, 9% of LPG-powered cars, 0.5% of hybrid cars and 0.5% of electric cars for the initial year (2019) and made by 25% of diesel-powered cars, 30% of petrol-powered cars, 20% of LPG-powered cars, 15% of hybrid cars and 10% of electric cars for the end year (2044*)

Note that the "CostParameters worksheet" of the LCCE tool provides default values for the unit costs to be used to calculate the costs related to delay, safety performance and emissions for each intersection alternative, and so on. LCCET uses the real discount rate to exclude the effects of expected inflation and to calculate annualized costs from net present costs. The discount rate was varied between 2% and 6%, but the values of 4–5% are the most representative values for investments in the public works sector [45]. Having set the value of the real discount rate, the calculation of the Net Present Values was made. Figure 5.10 shows the graphical results of the comparative analysis where reference is made to the discount rate of 5% and to the case where vehicle power engines varied within the fleet. Figure 5.10, indeed, refers the case where the traffic mix with 45% of diesel-powered cars, 45% of petrol-powered cars, 9% of LPG-powered cars, 0.5% of hybrid cars and 0.5% of electric cars is used for the initial year (2019), and 25% of diesel-powered cars, 30% of petrol-powered cars, 20% of LPG-powered cars, 15% of hybrid cars and 10% of electric cars is used for the end year (2044*). Benefit-Cost Ratios of 9.7 and 6.75 resulted for the alternative 1 (roundabout) and the alternative 2 (traffic signal), respectively; they were comparable to the values obtained when no change in vehicle power engines within the fleet was assumed.

Figure 5.11 illustrates the graphical results of the analysis only with reference to the cost items of emissions varying the vehicle power supply under two different vehicular mixes just for the final year (2044); reference is made to the discount rate of 5%. At last, Fig. 5.12 shows the variation of the net present value (NPV) of the design alternatives varying both the real discount rate and power characteristics of vehicles, that is using two different vehicular mixes for the end year (2044).

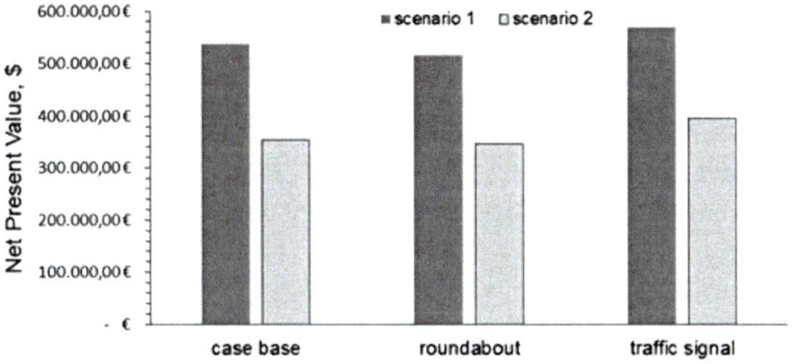

Fig. 5.11 Net present value of emission costs from intersection alternative analysis. *Note* Scenario 1 stands for a traffic mix made by 45% of diesel-powered cars, 45% of petrol-powered cars, 9% of LPG-powered cars, 0.5% of hybrid cars and 0.5% of electric cars for the initial year (2019) and the end year (2044); scenario 2 stands for a traffic mix made by 45% of diesel-powered cars, 45% of petrol-powered cars, 9% of LPG-powered cars, 0.5% of hybrid cars and 0.5% of electric cars for the initial year (2019) and made by 25% of diesel-powered cars, 30% of petrol-powered cars, 20% of LPG-powered cars, 15% of hybrid cars and 10% of electric cars for the end year (2044*)

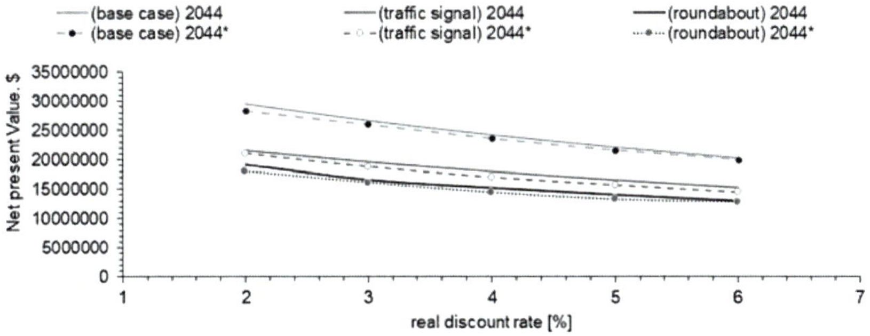

Fig. 5.12 Variation of the NPV varying the design alternatives and vehicle power supply. *Note* A traffic mix made by 45% of diesel-powered cars, 45% of petrol-powered cars, 9% of LPG-powered cars, 0.5% of hybrid cars and 0.5% of electric cars for the base year (2019) and the end year(2044); a traffic mix made by 45% of diesel-powered cars, 45% of petrol-powered cars, 9% of LPG-powered cars, 0.5% of hybrid cars and 0.5% of electric cars for the base year (2019) and made by 25% of diesel-powered cars, 30% of petrol-powered cars, 20% of LPG-powered cars, 15% of hybrid cars and 10% of electric cars for the end year (2044)

5.3 Discussion and Conclusions

The chapter introduces the Life-Cycle Costing decision-making methodology and presents a case study with the aim of evaluating the total life-cycle costs of three alternative intersection layouts. The Life-Cycle Cost Estimating Tool proposed by [38] has been applied to estimate the total net present values of benefits and costs over the life-cycle of each alternative intersection project. Some assumptions were made for the purposes of the case study and then 8 steps were followed to estimate the input data necessary to determine the cost items and Net Present Value for the design alternatives. Note that a reference period of 25 years from the 2019 to 2044 has been chosen to provide a common comparison point between the design alternatives. It was deemed consistent with environmental and geometric policies as discussed in [65]. This choice was also made in analogy to periods of analysis used for similar road infrastructure projects in the Italian context.

Another assumption concerned the two different composition of the vehicle fleet for the end year (2044). The reason for the dual choice was made to consider the progressive transition to electric vehicles, and to isolate and to test their effects on environment.

The analysis of findings shows that it is possible to claim that the roundabout alternative has many advantages than the other layouts during its useful lifetime in term of less delay time, and numbers of peak hours whatever maintenance program scenario. Concerning delays experienced by users, Fig. 5.6 shows that the roundabout (alternative 1) and the signalized intersection (alternative 2) resulted more efficient than the base case, having lower delay values both for the initial year and the end year. When it should be considered safety features in terms of expected number of crashes, the maintenance programs here considered are quite equivalent. However, the roundabout layout shows the highest performance in terms of road safety standards.

The summary of the road—maintenance strategies and costs in Table 5.8 highlighted higher costs for the design alternatives 1 to 2 than the case base. More specifically, the roundabout (alternative 1) is expected to have a cost greater than 7% compared to the base case, but a salvage value significantly higher than the base case (see Fig. 5.9); the signalized intersection (alternative 2) in turn, is expected to have a cost greater than 15% compared to the base case and a salvage value about twice the base case. The life-cycle cost evaluation shows that the roundabout alternative is characterized by the least cost in terms of net present value for the range of the discount rate values here considered. The comparison of emission values (see Fig. 5.8) shows that the roundabout layout provides less environmental pressure. Only the $PM_{2.5}$ emission values are very similar among the alternative solutions and these findings are related to the low speed profile associated with the road intersection operations. Furthermore, what it has been observed is strong linked to the fleet composition hypothesis and its variation during the road system lifetime. However, the hypothesis of growth of hybrid, electric and LPG-powered cars involved a reduction of about 30% in environmental costs compared to the scenario where the vehicle power supply unchanged (see Fig. 5.11). A comprehensive framework referring to mobility solutions for so-called smart city could provide many trends for innovative vehicles market penetration as well discussed in [66, 67]. Finally, considering the whole analysis period 2019–2044, the existing base alternative has the lowest salvage value under the maintenance program considered; then, the existing intersection layout is seemingly the lasting solution in terms of LCC value. The results confirmed the flexibility of the methodology also when environmental criteria should be integrated to conceptualize design solutions of urban intersections and to provide perspective on its performance to satisfy the major design needs starting from the planning stage.

Except for mobility investment decisions, bridge design, management of structural elements of road infrastructures, highway maintenance and rehabilitation strategies, the application of life-cycle costing methods for the construction projects especially at urban level is still fairly limited. The methodological approach showed for the application to road intersections could be easily extended in terms of road network assessment. In other words, the promise of circular city advantages needs to apply the LCC methodology at urban network level in order to provide both a powerful decision-making tool and a desirable policy of continuous improvement. In this view, the study presented in this chapter can contribute to fill the existing gap on the topic.

References

1. Kaewunruen S, Sussman JM, Matsumoto A (2016) Grand challenges in transportation and transit systems. Front Built Environ 2:4
2. Rosłon J, Książek-Nowak M, Nowak P, Zawistowski J (2020) Cash-flow schedules optimization within life cycle costing (LCC). Sustainability 12(19):8201
3. Petit-Boix A, Llorach-Massana P, Sanjuan-Delmas D, Sierra-Perez J, Vinyes E, Gabarrell X, Rieradevall J, Sanye-Mengual E (2017) Application of life cycle thinking towards sustainable cities: a review. J Cleaner Produc 166:939–951
4. Keoleian GA, Menerey D (1994) Sustainable development by design: review of life cycle design and related approaches. J Air Waste Manag Assoc 44(5):645–668
5. Kirchherr J, Reike D, Hekkert M (2017) Conceptualizing the circular economy: an analysis of 114 definitions. Resour Conserv Recycl 127:221–232
6. The road to circularity (2019). Why a Circular Economy is Becoming the New Normal, https://www.pwc.nl/en/assets/documents/pwc-the-road-to-circularity-en.pdf
7. Parikka-Alhola K, Nissinen A (2012) Environmental impacts of transport as award criteria in public road construction procurement. Int J Constr Manag 12:35–49
8. Thoft-Christensen P (2012) Infrastructures and Life-cycle cost-benefit analysis. Struct Infrastr Eng 8(5):507–516
9. Dhillon BS (1989) Life cycle costing: techniques. Models and Applications. Routledge, Amsterdam
10. Langdon D (2007) Towards a common european methodology for life-cycle costing (LCC): a common methodology final report—Guidance Document, Davis Langdon Management Consulting
11. Ratna RV, Kurian M, Ardakanian R (2015) Life-cycle cost approach for management of environmental resources: a primer. Springer Briefs in Environmental Science. Springer, Heidelberg
12. Wennström J (2014) Life cycle costing in road planning and management: a case study on collision-free roads. TRITA-BKN. Bulletin 125, 2014 KTH School of ABE ISSN 1103-4270
13. Prendeville S, Cherim E, Bocken N (2018) Circular cities: mapping six cities in transition. Environ Innov Societal Trans 26:171–194
14. Paiho S, Mäki E, Wessberg, N, Paavola M, Tuominen P, Antikainen M, Heikkilä J, Rozado CA, Jung N (2020) Towards circular cities—conceptualizing core aspects. Sustain Cities Soc 59(102143):1–19
15. European Investment Bank (2018). The 15 circular steps for cities. https://doi.org/10.2867/39283
16. Fusco Girard L, Nocca F (2019) Moving towards the circular economy/city model: which tools for operationalizing this model? Sustainability 11(22):6253
17. Marin J, De Meulder B (2018) Interpreting circularity. Circular city representations concealing transition drivers. Sustainability 10(5):1310

18. Alrawi F, Alwani K, Alacash H, Mesrop S (2021) Community participation towards sustainability enhancement of transportation sector for Baghdad City. In: Nathanail EG, Adamos G, Karakikes I (eds) Advances in mobility-as-a-service systems, vol 1278. AISC. Springer, Cham, pp 106–115
19. Danchuk V, Bakulich O, Svatko V (2019) Building optimal routes for cargo delivery in megacities. Transport and Telecommunication 20(2):142–152
20. Macioszek E, Sierpiński G, Staniek M (2017) Analysis of trends in development of freight transport logistics using the example of Silesian Province (Poland)—a case study. Transp Res Procedia 27:388–395
21. Sierpiński G, Macioszek E (2020) Equalising the Levels of Electromobility Implementation in Cities. In: Mikulski J (ed) Research and the future of telematics, vol 1289. Communications in computer and information science. Springer, Heidelberg, pp 165–176
22. Williams J (2019) Circular cities: challenges to implementing looping actions. Sustainability 11(2):423
23. Corona B, Shen L, Reike D, Rosales CJ, Worrell E (2019) Towards sustainable development through the circular economy—a review and critical assessment on current circularity metrics. Resour Conserv Recycl 151:
24. European Commission—Joint Research Centre—Institute for Environment and Sustainability: International Reference Life Cycle Data System (ILCD) Handbook—General guide for Life Cycle Assessment—Detailed guidance. First edition. Publications Office of the European Union, Luxembourg (2010)
25. Zamagni A, Pesonen H, Swarr T (2013) From LCA to life cycle sustainability assessment: concept, practice and future directions. Int J Life Cycle Assess 18:1637–1641
26. Abraham DM, Dickinson RJ (1998) Disposal costs for environmentally regulated facilities: LCC approach. J Constru Eng Manag 124(2):146–1 54
27. Heralova RS (2017) Life cycle costing as an important contribution to feasibility study in construction projects. Procedia Eng 196:565–570
28. Almeida JO, Delgado RM, Teixeira PF (2017) A bridge life-cycle cost optimization methodology. life-cycle of engineering systems: emphasis on sustainable civil infrastructure—Bakker, Frangopol & van Breugel (Eds), Taylor & Francis Group, London, pp 1103–1110
29. Klatter HE, van Noortwijk JM (2007) LCC analysis of structures on a network level in the netherlands. In: 5th International workshop on life-cycle cost analysis and design of civil infrastructure systems, pp 215–220, Hyo-Nam Cho, Dan M. Frangopol, Alfredo H-S Ang eds. Taylor & Francis, Seoul, Korea—London
30. Asadi P, Hajirasouliha I (2020) A practical methodology for optimum seismic design of rc frames for minimum damage and life-cycle cost. Eng Struct 202:
31. Xiao Y, Ni F, Dong Q (2008) Life-cycle cost optimization of highway maintenance and rehabilitation strategies based on integrated maintenance management system. In: 87th TRB annual meeting, transportation research board, paper 08–1376. Washington DC
32. Fuller SK, Petersen SR (1995) Life-cycle costing manual for the federal energy management program. NIST Handbook 135, MNW Publication
33. Goh KC, Yang J (2009) Developing a life-cycle costing analysis model for sustainability enhancement in road infrastructure project. rethinking sustainable development: planning, infrastructure engineering, design and managing urban infrastructure. queensland university of technology, Brisbane, Queensland
34. Brilon W (2000) Traffic flow analysis beyond traditional methods. In: Proceedings of the Fourth international symposium on highway capacity, transportation research board, Washington, DC, US, pp 26–41
35. Persad KR, Bansal S (2004) Toll road life-cycle: two case studies. In: 83rd TRB annual meeting. transportation research board, Washington DC, US
36. Federal Highway Administration (FHWA). Life-Cycle Cost Analysis. Real Cost user Manual, Department of Transportation Federal Highway Administration Office of Asset Management, Washington DC, US (2004)

37. Amini AA, Mashayekhi M, Ziari H, Nobakht S (2011) Life cycle cost comparison of highways with perpetual and conventional pavements. Int J Pavement Eng 13(6):1–16
38. National Academies of Sciences, Engineering, and Medicine. Estimating the Life-Cycle Cost of Intersection Designs. Washington, DC: The National Academies Press. https://doi.org/10.17226/21928 (2016)
39. UNEP-SETAC 2013. The Methodological Sheets for Subcategories in Social Life Cycle Assessment (S-LCA). UNEP-SETAC Life-cycle Initiative, Paris, France (2013)
40. UNEP-SETAC 2011. Towards a Life Cycle Sustainability Assessment: Making Informed Choices on Products. UNEP-SETAC Life-cycle Initiative, Paris, France (2011)
41. Highway Capacity Manual. Highway Capacity Manual, 5th edition. Washington, DC: Transportation Research Board 1650 (2010)
42. Highway Safety Manual, 1st edition. American Association of State Highway and Transportation Officials, Washington DC, US (2010)
43. American Association of State Highway and Transportation Officials (2010) User and non-user benefit analysis for highways (Redbook). AASHTO, Washington, D.C
44. Pulugurtha SS, Penmetsa P, Duddu VR (2015) Monetizing reliability to evaluate the impact of transportation alternatives. NCDOT Project 2015–07 FHWA/NC/2015-07 February 2017, Final report, Raleigh, North Carolina, US
45. Guide to Cost-Benefit Analysis of Investment Projects Economic appraisal tool for Cohesion Policy 2014–2020 European Commission Directorate-General for Regional and Urban policy REGIO DG 02, https://doi.org/10.2776/97516. https://ec.europa.eu/regional_policy/sources/docgener/studies/pdf/cba_guide.pdf
46. Automobile Club of Italy. Car trend 2012–2018, statistical analysis on car-market trends in Italy, Statistic area, studies & searches direction. Management Control office. Automobile Club of Italy & National Institute of Statistics, Rome, Italy (2018)
47. Lo Pinto A (2019) Assessing the life-cycle costs for road intersection projects. Master Degree Thesis in Civil Engineering, University of Palermo, Italy
48. Italian Minister of Infrastructure Transport. (2006) Available online at: http://www.mit.gov.it/mit/site.php?p=normativa&o=vd&id=1735&idcat=&iddett=0
49. American Association of State Highway and Transportation Officials. AASHTO Guide for design of pavement structures (1993)
50. Huang YH (2003) Pavement analysis and design II ed., Pearson Education, Inc., Upple Saddle River, NJ 0745 8
51. AASHTO Pavement Management Guide (2012)
52. Sicily Region, Regional Department of Infrastructures and Mobility, Single Regional Price List for LL. PP. (2019)
53. National Academies of Sciences (2010) Engineering, and medicine, roundabouts: an informational guide, 2nd edn. The National Academies Press, Washington, DC, USA
54. European Commission, https://ec.europa.eu/programmes/horizon2020/en/h2020-section/smart-green-and-integrated-transport
55. Technical Support Document: Technical Update of the Social Cost of Carbon for Regulatory Impact Analysis Under Executive Order 12866 (May 2013; revised November (2013)
56. Corporate Average Fuel Economy for MY2017- MY2025 Passenger Cars and Light Trucks (August 2012), page 922, Table VIII-16, "Economic Values Used for Benefits Computations" (2010)
57. Transport Innovation Software (TIS), free to downloaded at http://tis.unical.it/index.php/development/tritone/
58. Panis LI, Broekx S, Liu R (2006) Modeling instantaneous traffic emission and the influence of traffic speed limits. Sci Total Environ 371(1–3):270–285
59. Witzig R (2010) The role of electric vehicles in achieving climate goals. In: European Transport Conference, pp. 1–14. Glasgow, United Kingdom. The Association for European Transport
60. Transport & Environment, https://www.transportenvironment.org/news/does-electric-vehicle-emit-less-petrol-or-diesel

61. Capros P, De Vita A, Tasios N, Siskos P, Kannavou M, Petropoulos A et al (2016) Energy, transport and GHG emissions -trends to 2050. Eur Community, p 27
62. Barceló J (2010) Fundamentals of Traffic Simulation. Springer, London
63. Euro missions Standards Limits to Improve Air Quality and Health, https://www.theaa.com/driving-advice/fuels-environment/euro-emissions-standards
64. Curry DA, Anderson DG (1973) Procedures for estimating highway user costs, air pollution, and noise effects. NCHRP Report 133, Transportation Research Board
65. http://onlinepubs.trb.org/Onlinepubs/nchrp/nchrp_rpt_133.pdf
66. Fernandes P, Tomás R, Acuto F, Pascale A, Bahmankhah B, Guarnaccia C, Granà A, Coelho MC (2020) Impacts of roundabouts in suburban areas on congestion-specific vehicle speed profiles, pollutant and noise emissions: An empirical analysis. Sustain Cities Soc 62, 102386
67. Šurdonja S, Giuffrè T, Deluka-Tibljaš A (2020) Smart mobility solutions-necessary precondition for a well-functioning smart city. Transp Res Procedia 45:604–611

Chapter 6
Method of Assessing the Impact of the Socio-financial Conditions on the Bike-Sharing System Operation and Its Implementation in Medium-Sized Cities

Agnieszka Tubis, Emilia Skupień, and Mateusz Rydlewski

6.1 Introduction

The pioneer in the study of the sharing economy phenomenon is Lawrence Lessig, who called in this words the consumption implemented by sharing, exchange and lending one's own resources without transferring ownership of the goods [1]. Sharing economy has become popular due to the inhabitants' knowledge about IT techniques and their readiness for interactivity. Generation Y, so-called Millennials, are looking at the world differently, and they adapt and perfectly understand peer-to-peer systems. In addition, raising the awareness of residents and striving to limit consumerism means that they need and they are looking for alternatives to their properties, and thus are open to sharing economy [2]. The internet platforms are an important element connecting the sharing economy and the city, allowing to share large amounts of data. Thanks to the possibilities of offering access to large data sets, cities can offer a wide range of services tailored to the expectations of residents. Because of them, sharing mobility can be developed [3]. Sharing economy is increasingly used in various areas of human activity. The areas in which the economy of sharing is the most developed include: transport, storage, tourism, housing, food, media, free time [4, 5]. According to research carried out by PWC (Pricewaterhouse Coopers), in the year 2025 revenues from consumption based on access will increase to USD 300 billion [6].

A. Tubis (✉) · E. Skupień · M. Rydlewski
Wroclaw University of Science and Technology, Wrocław, Poland
e-mail: agnieszka.tubis@pwr.edu.pl

E. Skupień
e-mail: emilia.skupien@pwr.edu.pl

M. Rydlewski
e-mail: mateusz.rydlewski@pwr.edu.pl

E. Macioszek and G. Sierpiński (eds.), *Research Methods in Modern Urban Transportation Systems and Networks*, Lecture Notes in Networks and Systems 207, https://doi.org/10.1007/978-3-030-71708-7_6

Authors [7] in their chapter stated that the sharing economy is not really a 'sharing' economy at all; it is an access economy. According to the authors, sharing sensu stricto is a form of social exchange that occurs between friends and does not bring any profit. If, however, the sharing takes place with the participation of the market, when the company becomes an intermediary between consumers who do not know each other, this practice is based on paid access to the other party's goods and services within a certain time and is in the nature of economic exchange. In this list, clients obtain utility value rather than social value, in which case we are talking about the access economy. Thus, access-based consumption is usually defined as a transaction that can take place via the market, and in which there is no transfer of ownership [7]. The right of ownership remains with the supplier who bears the related burdens, such as, for example, liability for repairs [8]. The consumer acquires or obtains the right to use the product for a certain period of time, for which he charges a fee when the market mediation takes place [9]. Transport is a field in which the sharing economy develops very dynamically, both in the C2C and B2C relations [10]. Urban transport based on the access economy concept is one of the basic instruments by which cities become more intelligent [11–13].

Bike-sharing is one of the forms of urban transport based on the access economy idea. Four generations of public bike-sharing systems (PBS) are historically distinguished in literature [14]. The standard first-generation system is called the White Bicycle system, which originated in Amsterdam in 1965. The plan was to collect old bikes, paint them white and make them available for free in the city. However, the system did not succeed as a result of opposition from the authorities and acts of vandalism, but the idea itself appeared in the minds of residents [15]. The introduction of city bikes in Copenhagen in 1995 is considered as the beginning of the second generation systems [16]. The system involved a fee and worked on a coin-deposit principle. Assumptions of this system have not changed in the next generations of PBS (distinctive design and colour of bicycles, special docking stations serving the basic operations on rent and return, paid rent). Experience in the operation of second-generation systems allowed gaining experience, which is used for systems of the third generation. The introduction of BIXI in Montreal in 2009 [17] should be considered as the beginning of the third-generation systems. BIXI offers new solutions and concepts for both technical and organisational sides, which became the basis for the third generation of PBS in Canada and the United States [18, 19]. Basic features of this generation are the integration between transport and advertising functions, additional features of the docking station (user identification and payment service) and the use of advanced information technology (mobile phones, magnetic stripe cards, smartcards). Most currently operating PBS belong to the third generation. Some authors [14] postulate the separation of the fourth generation of PBS. It contains all of the features particular to the third-generation systems, but also aims for the integration of cycling with other modes of public transport (especially car-sharing). The basic method of integration is common in determining the locations of stations and stops for all modes of transport. It also increases the use of advanced technological solutions (solar systems and bicycles with electric drive in order to promote the principles of sustainable development) [16].

The public bike-sharing system is also intensively developed in cities of Poland. The first city to implement this solution was Krakow in 2008. Currently, bike-sharing already operates in all large cities with over 250,000 and shows potential for further development. Searching for new markets for bike-sharing services, their suppliers are looking with increasing interest in medium-sized cities. However, the specificity of the regional and sub-regional city means that the implementation rules and the target user segment may differ from the traditionally described solutions. The aim of the chapter is to present the method of assessing the impact of socio-financial conditions on the implementation and functioning of bike-sharing systems in medium-sized cities and the results obtained in the study of selected Polish cities. To achieve this goal, subchapter 2 presents the bike-sharing system in Poland, which was developed in large cities. Subchapter 3 describes the proposed research method, focused on the specificity of the regional and sub-regional city. On its basis, the following subchapters present the results of the analyses and the conclusions made on their basis. The last subchapter presents the final conclusions.

6.2 Bike-Sharing in Large Cities in Poland

In Poland, a bike-sharing boom has been observed for several years. The positive effects of the functioning of the PBS in large cities related to the reduction of the level of pollution place it in the assumptions of the concept of sustainable urban transport [20–22] and intelligent mobility [23]. Supporting the development of infrastructure for cyclists is not only a positive action for the benefit of two-wheelers, but also for the image of the city [24], by creating it as open to environmentally friendly solutions [25]. This solution significantly improves the quality of transport in the city by reducing pressure on the environment [26], as well as promoting sustainable urban transport [27, 28], especially cycling [29].

In Poland, as in other European countries, in the first place the bike-sharing boom covered mainly large cities with over 250,000 inhabitants—the implementation of other cities is shown in the Table 6.1.

Residents of large urban agglomerations use bike-sharing systems that are still under development with great enthusiasm. This is mainly due to:

- greater ecological awareness—residents of large cities more often experience the effects of motorized urban traffic (smog, noise),
- extended travel time on the route to/from work, caused by congestion and heavy urban traffic,
- large distances separating the place of residence from the place of work (and other destinations), preventing the movement only on foot,
- paid parking zones and a shortage of parking spots, in particular in the city centre,
- greater willingness to use mobile applications in everyday life, and in particular in handling travels carried out in urban space.

Table 6.1 FBS implementations in major cities in Poland

City	Number of residents	Year of PBS start
Warszawa	1,769,529	2012
Kraków	769,498	2008
Łódź	687,702	2016
Wrocław	639,258	2011
Poznań	537,643	2012
Gdańsk	464,829	2019
Szczecin	403,274	2014
Bydgoszcz	355,006	2016
Lublin	339,811	2014
Białystok	297,403	2014
Katowice	295,449	2015

The authorities of large cities are also very interested in the development of alternative means of transport that will limit urban traffic, which will increase the quality of life of residents and improve the image of the city itself and its tourist attractiveness. For this reason, in addition to the implementation and development of the bike-sharing system itself, the authorities of large cities are also increasingly investing in infrastructure accompanying cycling systems. Thanks to this, one can observe a growing number of bicycle paths built in large cities of Poland. This can be confirmed by the results of the Bicycle Ranking of Polish Cities [30].

The bike-sharing boom observed in Poland is primarily influenced by four groups of factors: political, technological, social and environmental (Fig. 6.1).

The presented factors cause a steady increase not only in new city bike rentals in other cities, also medium-sized ones, but also the development of already functioning systems in Polish cities (Figs. 6.2 and 6.3).

This development is certainly influenced by the regularly increasing number of rentals (Fig. 6.4) by regular users of this system, as well as the increasing number of new interested cyclists (Fig. 6.5).

At the same time, the characteristics of the main PBS users in Poland in large cities correspond to the results of studies conducted in the United States [32]. They are primarily people with significantly higher employment rates and education levels, lower average age, and more likely to be male. Also the results of the research by Fuller et al. [33] confirmed that members of the BIXI program in Montreal to be skewed toward the 18–24 years band. They also found users to be more likely to have a tertiary education. At the same time, according to the research presented in [14] bike share users in North America were found to be more likely, than the general population, to live closer to their work. A similar tendency can also be observed in large cities in Poland.

Research conducted in China [34] and in the United States [14, 32], as well as others countries [35, 36] indicate that users of the bike-sharing system use vehicles

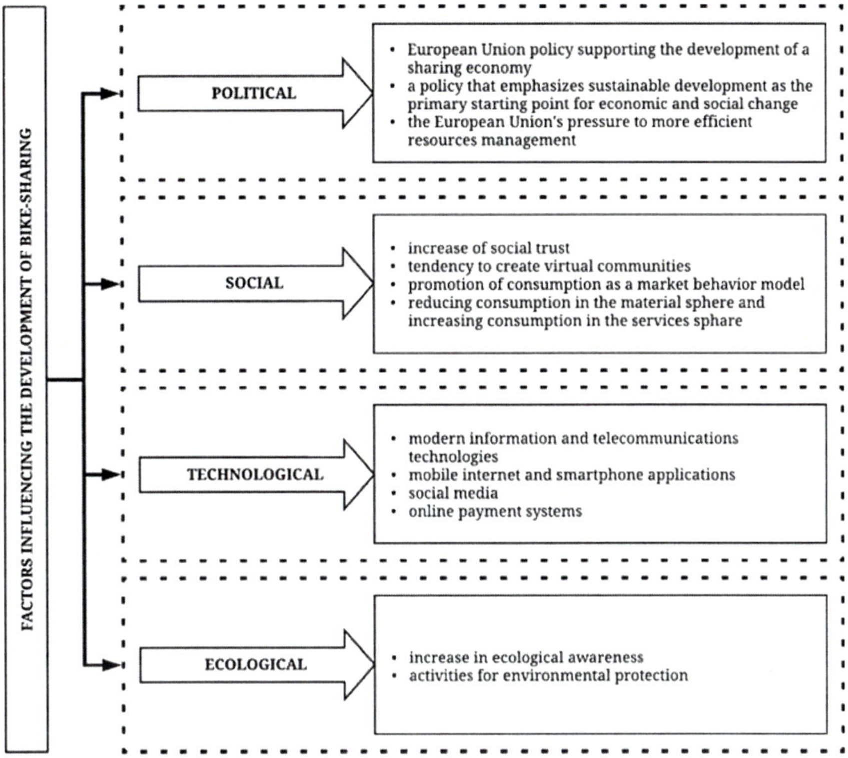

Fig. 6.1 Factors influencing the development of bike-sharing. *Source* Own works based on [5]

Fig. 6.2 Number of PBS rental stations. *Source* Nextbike Polska [31]

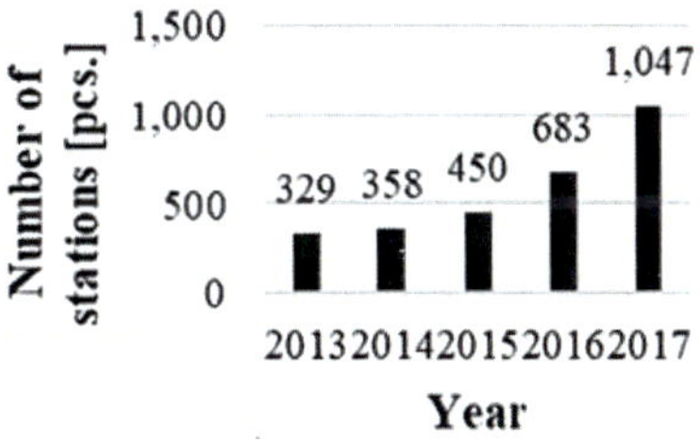

Fig. 6.3 Number of bikes of PBF. *Source* Nextbike Polska [31]

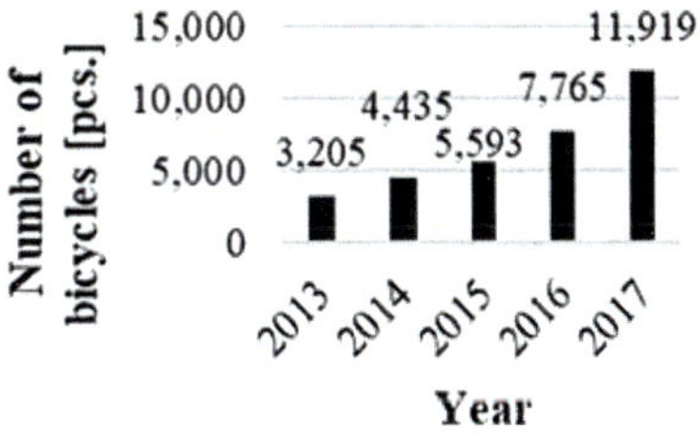

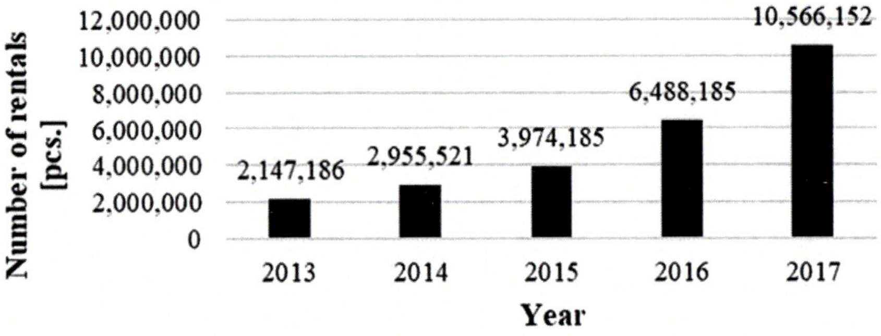

Fig. 6.4 Number of rentals. *Source* Nextbike Polska [31]

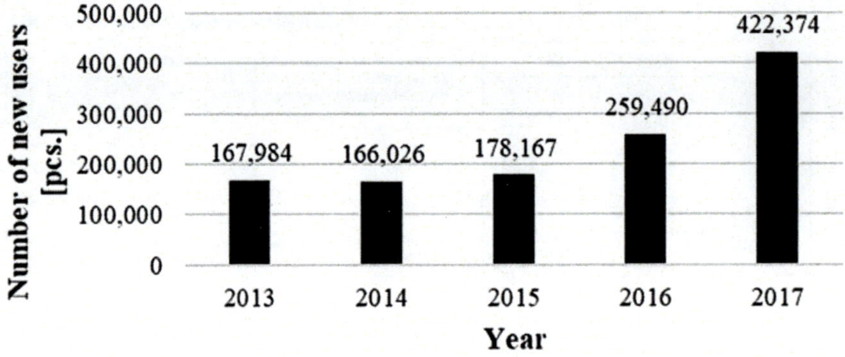

Fig. 6.5 Number of new users. *Source* Nextbike Polska [31]

primarily for commuting to/from work/school and for social gatherings. The results of the users of the Wroclaw City Bike confirmed similar tendencies also in Polish big cities [2].

6.3 Proposed Research Method

In the conducted research, authors used the method of assessing the impact of socio-financial conditions on the implementation and functioning of the bike-sharing system in a selected group of cities. This method involves 6 steps presented in the Fig. 6.6.

Step 1: Based on the literature review, social and financial factors that favour the development of the bike-sharing system have been identified. Literature review and some existing evidence suggests that users of public bike-sharing schemes, like cyclists in general, tend to have higher incomes, high levels of formal education, and be disproportionately white, middle aged and male. At the same time, due to the fact

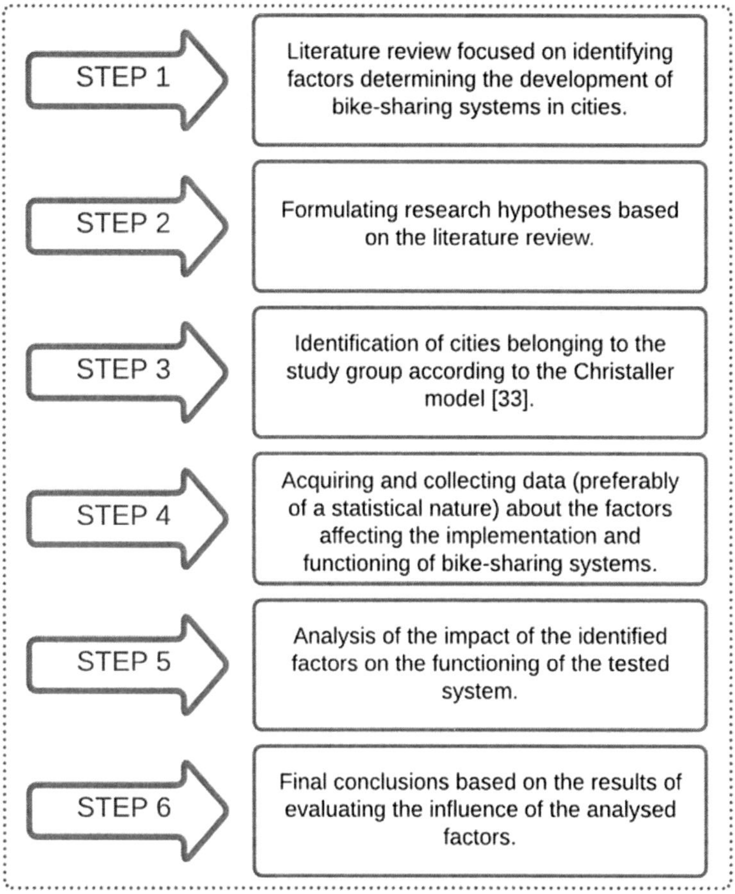

Fig. 6.6 6 steps of method

that in many cases, cities participate in the costs of functioning of the bike-sharing system, this solution is implemented in cities with a high budget. For this reason, an important factor under assessment may be the average income per capita.

Step 2: Based on the implementation of Step 1, it was possible to formulate 3 research hypotheses:

H1: Bike-sharing systems are created in cities with a large share of people in the so-called mobile age among residents, with the mobile age range of 18–45 years.

H2: Bike-sharing systems are created in cities with sufficiently high incomes, in which the share of collective transport costs constitutes a significant budget item.

H3: Bike-sharing systems are created in cities whose income per capita is above the average.

The hypotheses made do not take into account the skin colour of inhabitants of medium-sized cities due to the dominant share of white people in Poland.

Step 3: Based on Christaller's model related to the structure of the urban settlement patterns in Poland [37], two groups of medium-sized cities occurring in Poland were distinguished: (1) regional city −100,000–250,000 inhabitants and (2) sub-regional city −50,000–100,000 inhabitants. Therefore, 73 cities were accepted for analysis, 28 of which were included in the regional city group and 45 in the sub-regional city group.

In order to verify the hypotheses, 10 city representatives in both distinguished groups were selected for detailed analysis. In order to maintain the ratio of cities with a bike-sharing system to those that have not yet implemented this solution, the following detailed analysis was adopted:

- in the regional city group −7 cities with bike-sharing and 3 cities without the system,
- in the sub-regional city group −3 cities with bike-sharing and 7 cities without the system.

Step 4 and step 5: The first stage of the analysis concerned the entire studied population of cities. The research was based on data obtained from city offices and the Central Statistical Office, as well as from the Public Information Bulletin. Detailed studies of city representatives included analyses of statistical data on:

- demographic measurements covering the number of inhabitants in selected age groups,
- analysis of average earnings per capita in the city.

Information on the education of the inhabitants was not included in the analysis because of the lack of reliable data in this respect. Based on the data from the Public Information Bulletin and city offices, income and costs budgeted by the authorities of the surveyed representatives were analysed. The detailed analysis included:

- obtained general income in 2016–2019,
- average income per capita calculated on the basis of registered incomes in 2016–2019,
- overheads incurred in 2016–2019,
- costs incurred for urban transport in 2016–2019.

6.4 Bike-Sharing Systems in Medium-Sized Cities in Poland

Bike-sharing systems are more often implemented and more developed in large cities than medium-sized ones. This is largely due to the characteristics of the average user of this system, which was presented in point 2, as well as the specificity of urban transport systems and the potential of medium-sized cities. One can also observe

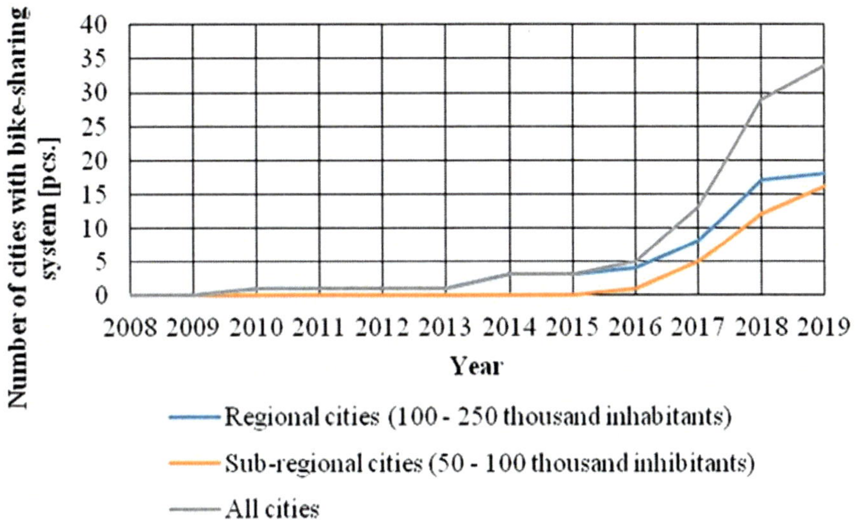

Fig. 6.7 The growth of cities having a bike-sharing system in the years 2008–2019

similar trends in Poland. However, some exceptions can be registered from this trend. Rzeszow (190,000 inhabitants) in 2010 initiated the first implementation of a public bicycle system in a medium-sized city and was the second city in Poland in which this system was implemented. In this way, it was even ahead of the country's capital (Warsaw implementation in 2012).

An analysis of the implementation of bike-sharing systems in Polish medium-sized cities (both regional and sub-regional) shows that the bike-sharing boom in this group is beginning—Fig. 6.7.

As can be seen from the Fig. 6.7 the interest in implementing the bike-sharing system in medium-sized cities can only be observed from 2017. It should be noted, however, that this applies to both regional and sub-regional cities. The number of cities with a public bicycle system is growing in the next two years, with the growth rate being higher in sub-regional cities, but this is also partly due to the fact that this group of cities is more numerous. Comparing both groups to each other, it should be clearly indicated that the regional city group with the implemented bike-sharing system is more numerous than the analogous sub-regional city group. This also translates into an increased percentage of cities with PBS in the regional city group, as shown in Fig. 6.8.

The share of cities with implemented PBS in the sub-regional group is smaller, but it should be noted that (1) the number of cities in this group is almost twice as large and (2) the implementation of PBS began later than in the regional city—the first full implementation took place in 2017.

The idea of a public city bike assumes that the city participates in the costs of operating the system. An analysis of the solutions implemented so far in both groups of medium-sized cities indicates that city authorities usually adopt one of three models of settlement of participation in the costs of operating PBS:

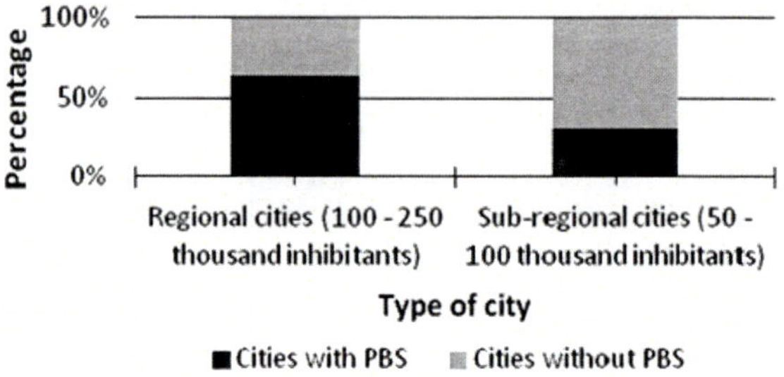

Fig. 6.8 Shares of cities with PBS system in the regional and sub-regional city

Table 6.2 Expenditure incurred by cities on PBS in 2019 highlighted as a separate budget line

A group of cities	Average costs on PBS	Maximum costs on PBS in the group	Minimum costs on PBS in the group
Regional	1,261,453 PLN	3,944,958 PLN	126,000 PLN
Sub-regional	172,656 PLN	641,000 PLN	800 PLN

- the city plans expenses in the form of clearly identified cost position in the budget allocated to PBS,
- the city plans expenses in the form of a cost item in the budget for all urban transport, in which the costs for PBS are one of the many components,
- the city transfers supervision over PBS to its subsidiary and the costs of operating PBS are part of the budget reported by the unit.

The analysis of budget data provided by cities, which distinguish expenditure on PBS as a separate budget item in their cost structure, indicates a large dispersion of these expenditures incurred in both analysed groups. The results of the analysis for 2019 are presented in the Table 6.2.

Expenditures on PBS in regional cities are of course higher than in the case of sub-regional cities. It should be noted, however, that in some sub-regional cities these outlays are much higher than in selected cities from the regional group. This can be seen even on the basis of a comparison of the average cost incurred on PBS in sub-regional cities compared to the minimum costs incurred by cities from the regional group.

6.5 Budget Analysis of Selected Medium-Sized Cities in Poland

The first hypothesis being verified concerns the share of mobile age residents living in given cities. The mobile age was assumed to be between 18 and 45 years old. Figure 6.9 shows the share of mobile age residents in a regional city by city with or without PBS, while Fig. 6.10 shows cities in the same division but from the sub-regional city group.

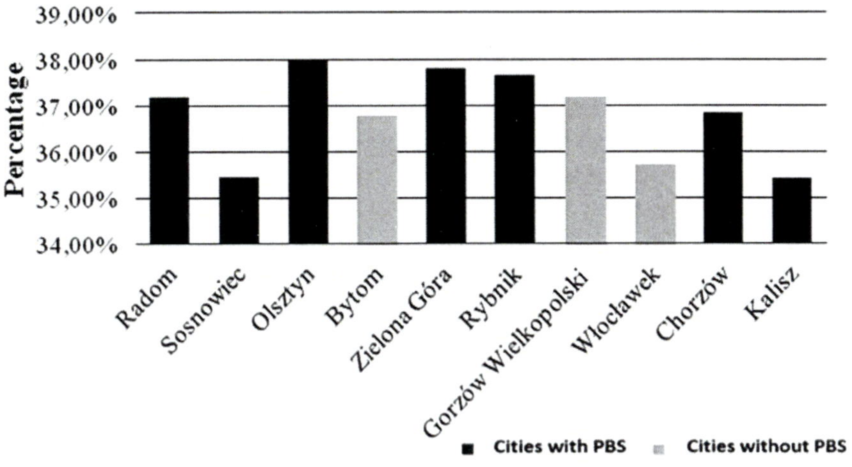

Fig. 6.9 Share of mobile age residents in regional cities

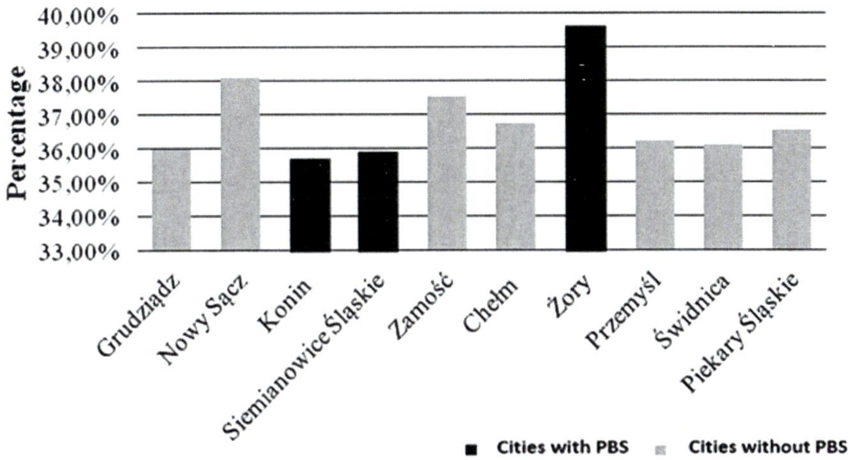

Fig. 6.10 Share of mobile age residents in sub-regional cities

As can be seen in the graphs presented, the increased participation of mobile-age residents has no significant impact on the implementation of the PBS system in medium-sized cities. This can be seen in both regional cities and sub-regional cities. In the sub-regional cities group, PBS have implemented cities whose percentage of mobile age residents is the lowest in the studied group, while cities with a higher share of mobile age residents have still not implemented this solution.

Based on Figs. 6.11 and 6.12, it can be seen that the average income from 2016–2019 achieved by cities from the studied group was not the basic premise for making decisions regarding the implementation of PBS. Both in the regional and sub-regional group there are cities that achieve income above the average for a given group and

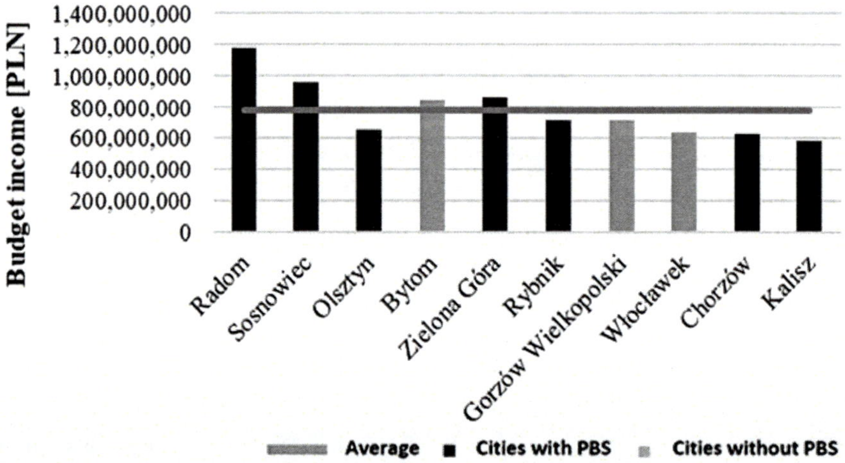

Fig. 6.11 Average budget income in regional cities

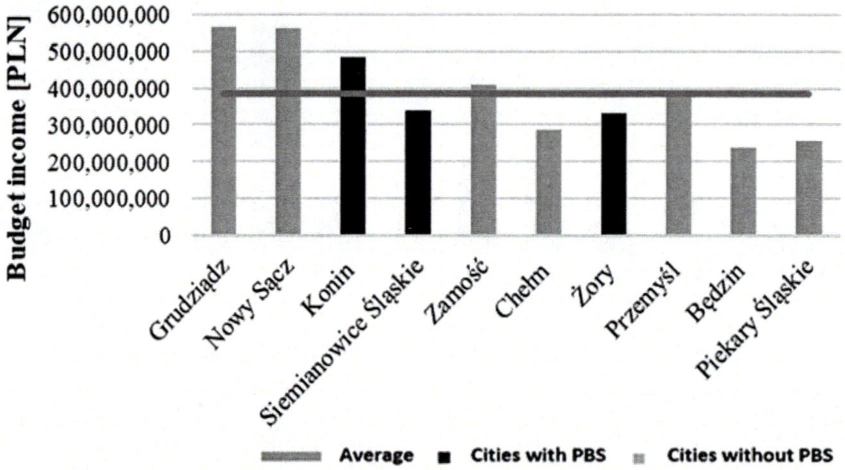

Fig. 6.12 Average budget income in sub-regional cities

do not have PBS implemented. In this case, it is noteworthy that two cities from the sub-regional group, with the highest average income, have not implemented PBS. At the same time, cities whose income is below average in the study group implemented bike sharing.

Hypothesis 2 is also not confirmed in the analysis of the costs incurred for public transport by cities in both examined groups. As shown in Figs. 6.13 and 6.14, the PBS system is implemented in cities from the regional and sub-regional groups whose share of transport expenditure is the lowest in the studied group. At the same time,

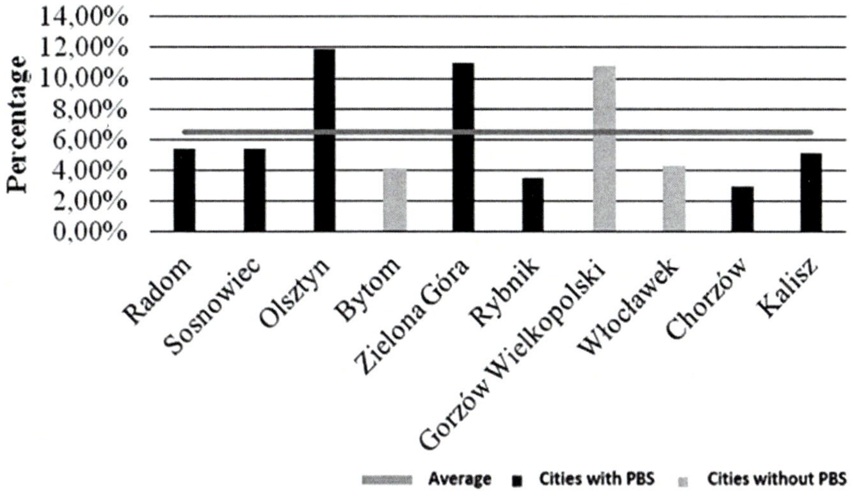

Fig. 6.13 Ratio of average costs of functioning public transport to average budget expenditure in regional cities

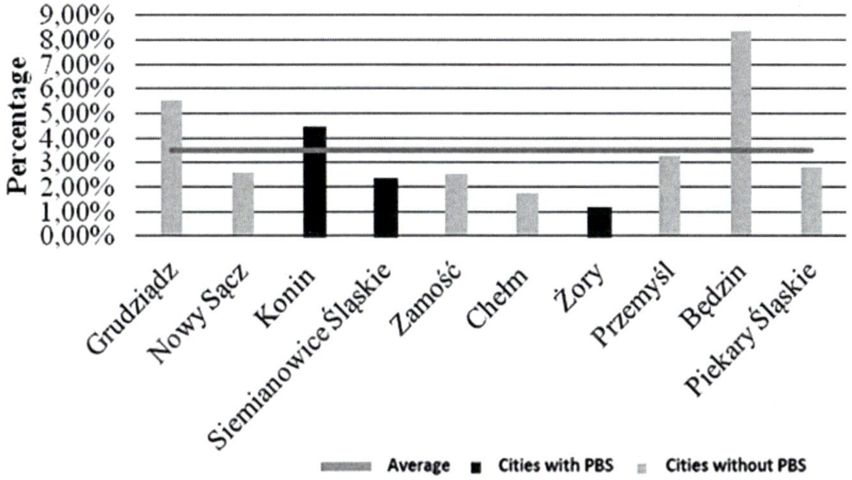

Fig. 6.14 Ratio of average costs of functioning public transport to average budget expenditure in sub-regional cities

there is no PBS in cities whose expenses for public transport are above the average share of transport costs in the total costs of cities calculated for a given group.

The cities accepted for analysis, both in the region and sub-regional group, differ in the number of inhabitants. For this reason, the results presented so far require confirmation in the form of an analysis of the city's income per capita. The results of the analysis are presented in Figs. 6.15 and 6.16.

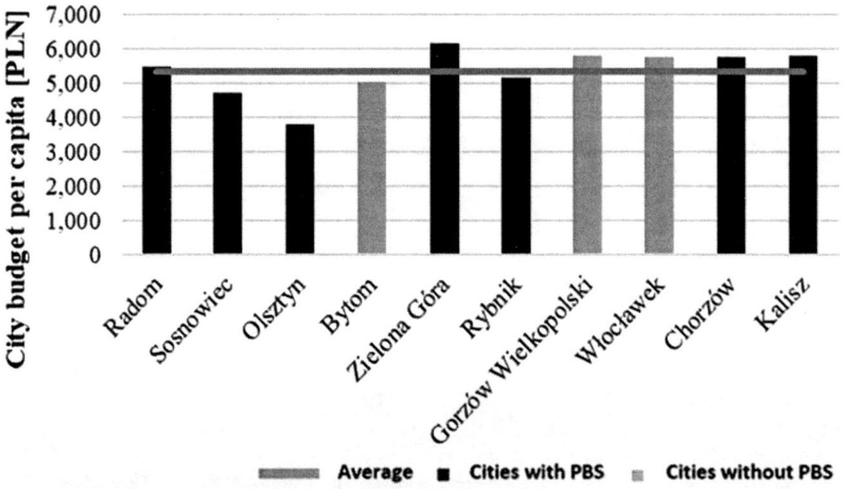

Fig. 6.15 Average city budget per capita in regional city

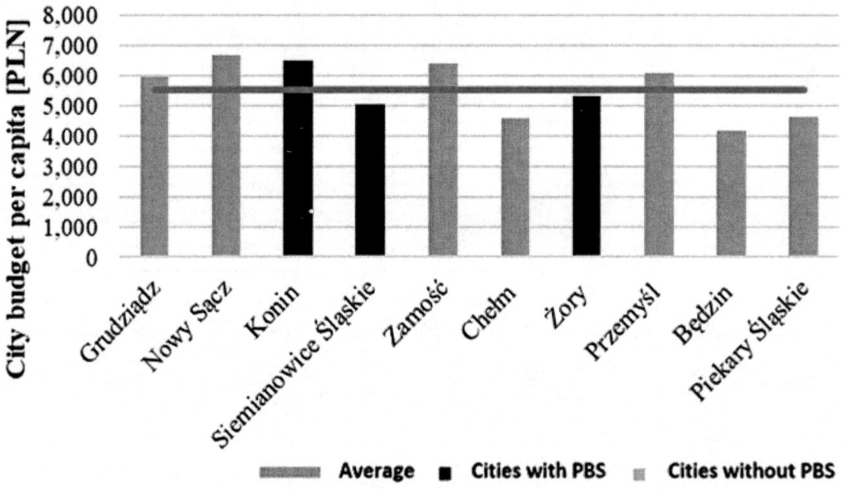

Fig. 6.16 Average city budget per capita in sub-regional city

The results of the average city's income per capita confirm the current conclusions. In both analysed groups of cities, average income per capita was not a reason for the decision regarding the implementation of PBS. For the regional city group, the city with the lowest per capita income (Olsztyn) already has PBS. While two cities (Gorzow Wielkopolski and Wloclawek), whose average income per capita is above the average income calculated for the group, still have not implemented PBS. In the sub-regional city group, 2 of the 3 cities that currently have PBS record the average income per capita in the last 4 years below the average income determined for this group of cities. Thus, hypothesis 3 set out in point 3 of this chapter has not been confirmed.

6.6 Conclusions

The research based on the presented method focused on proving 3 hypotheses formulated in Step 2 of the Method. Public bike-sharing systems were primarily dominated by large cities in Poland. However, in the last 3 years they are increasingly being implemented in regional cities with 100,000–250,000 inhabitants and in sub-regional cities −50,000–100,000 inhabitants. The reasons for implementing PBS solutions in medium-sized cities seem to be economic considerations related to the city's income, as well as environmental awareness expressed in the amount of expenditure on public transport. An additional demographic parameter that may influence the authorities' decision to admit such a system is the share of the number of inhabitants in mobile age. Therefore, the analyses carried out were aimed at proving 3 hypotheses related to the above assumptions.

The conducted research proceedings, based on data available from regional and sub-regional cities, as well as CSO statistical data, undermined the adopted hypotheses. Analyses of data from 2016–2019 from 10 regional and 10 sub-regional cities do not confirm the adopted relationships. The premise for the implementation of PBS for the analysed cities is neither the general income achieved by the city nor the income per capita, nor the share of transport costs in the total costs recorded by the city. This means that the premises for implementation should be sought in other areas, not necessarily quantitative. They may constitute the current trends in the development of urban transport, pressure from residents, lobbies of selected business entities or social organizations.

For this reason, the continuation of the research described in the chapter will be further analyses aimed at seeking premises prompting medium-sized cities in Poland to implement PBS. These studies will be aimed at identifying both quantitative and qualitative premises, which will significantly expand the scope of analyses. A benchmarking analysis is also foreseen to compare system solutions implemented in Polish cities with cities in other regions of the world.

Although the hypotheses pointed out in Step 2, have been disproved, the presented method can be used to analyze other data, e.g. on the factors influencing the implementation of car-sharing systems. Carrying out the procedures of the presented method leads to the formulation of hypotheses on the basis of the literature review. The hypotheses verification ultimately leads to a deeper understanding of the analyzed process.

References

1. Lessig L (2008) Remix: making art and commerce thrive in the hybrid economy. Penguin Press, New York
2. Voznakowa I, Kauf S (2018) Public transport in the era of automatisation and sharing economy. Proceeding of the Carpathian Logistics Congres 2018, Prague
3. Martin E, Shaheen S (2016) an analysis of five north american cities. impacts of cargo on vehicle ownership, modal shift, vehicle miles traveled, and greenhouse gas emissions. http://innovativemobility.org/wp-content/uploads/2016/07/Impactsofcar2go_FiveCities_2016.pdf. Access 20 August 2019
4. Collaborative Finance: The Sharing Economy. http://www.collaborativefinance.org/sharing-economy. Access 4 May 2017
5. Macioszek E (2019) E-mobility infrastructure in the Górnośląsko—Zagłębiowska Metropolis, Poland, and potential for development. In: Proceedings of the 5th world congress on new technologies. Paper No. ICERT 108, pp ICERT 108-1-ICERT 108-4. Lisbon
6. Tkaczyk J, Awdziej M (2017) Consumer motivations and attitudes toward car-sharing. Res Paper of Wroclaw Univ Econ 501:165–172
7. Bardhi F, Eckhardt GM (2012) Access-based consumption: the case of car sharing. J Consum Res. 39(4):881–898
8. Schaefers T, Lawson SJ, Kukar-Kinney M (2016) How the burdens of ownership promote usage of access—based services. Market Lett 27:569–577
9. Durgee F, O'Connor GC (1995) An exploration into renting as consumption behavior. Psychol Market 12(2):89–104
10. Koźlak A (2017) Sharing-economy as a new socio-economic trend. Res Paper Wroclaw Univ Econ 489:171–182
11. Kauf S (2018) Ekonomia współdzielenia (sharing economy) jako narzędzie kreowania smart city. Zeszyty Naukowe Politechniki Śląskiej, Seria: Organizacja i Zarządzanie 120:141–151
12. Macioszek E, Sierpiński G, Staniek M (2017) Analysis of trends in development of freight transport logistics using the example of Silesian Province (Poland)—a case study. Transp Res Procedia 27:388–395
13. Sierpiński G, Macioszek E (2020) Equalising the Levels of Electromobility Implementation in Cities. In: Mikulski J (ed) Research and the future of telematics, vol 1289. Commun Comput Inform Sci. Springer, Heidelberg, pp 165–176
14. Shaheen S, Martin E, Cohen AP, Finson R (2012) Public bikesharing in North America: Early operator and user understanding. Mineta Transportation Institute, San Jose
15. Jankowska-Karpa D, Wnuk A (2014) System roweru publicznego (Bike Sharing System—BSS) jako element polityki zrównoważonej mobilności na przykładzie Francji i Polski. Logistyka 6:4803–4812
16. Dobrzyńska E, Dobrzyński M (2016) Structure and dynamics of a public bike-sharing system. Case study of the public transport system in Białystok. Econ Manag 8 4):59–66
17. Faghih-Imani A, Eluru N, El-Geneidy AM, Rabbat M, Haq U (2014) How land-use and urban form impact bicycle flows: evidence from the bicyclesharing system (BIXI) in Montreal. J Tran Geo 41:306–314
18. Ahillen M, Mateo-Babiano D, Corcoran J (2016) Dynamics of bike sharing in Washington, DC and Brisbane, Australia: implications for policy and planning. Int J Sust Trans 10(5):441–454
19. Mahmoud MS, El-Assi W, Habib KN (2015) Effects of Built environment and weather on bike sharing demand: station level analysis of commercial bike sharing in Toronto. Trans Res Board 94th Ann Meet. Retrieved from http://docs.trb.org/prp/15-2001.pdf. Access 10 August 2019
20. Behrendt F (2016) Why cycling matters for smart cities. Int Bicycles Intell Transp. J Trans Geo 56:157–164
21. Midgley P (2009) The role of smart bike-sharing systems in urban mobility. Journeys 2(2):23–31
22. Mugion RG, Toni M, Raharjo H, Di Pietro L, Sebathu SP (2018) Does the service quality of urban public transport enhance sustainable mobility? J Cleaner Product 174:1566–1587

23. Benevolo C, Dameri RP, d'Auria B (2016) Smart Mobility in Smart City. In: Torre T, Braccini A, Spinelli R (eds) Empowering organizations, vol 11. Lecture Notes in Information Systems and Organisation. Springer, Cham, pp 13–28
24. Alexandros N, Pontus W, Oskar R (2016) The paradox of public acceptance of bike sharing in Gothenburg. Proc Inst Civil Eng Eng Sust 169(3):101–113
25. Gast N, Massonnet G, Reijsbergen D (2015) probabilistic forecasts of bike-sharing systems for journey planning. Proceeding of the 24th ACM international conference on information and knowledge management (CIKM 2015), Oct 2015, Melbourne, Australia
26. DeMaio P (2009) Bike-sharing: history, impacts, models of provision, and future. J Pub Trans 12(4):41–56
27. Bordagaray M, Ibeas A, dell'Olio L (2012) Modeling user perception of public bicycle services. Soc Beh Sci 54:1308–1316
28. Fishman E, Washington S, Haworth N (2014) Bike share's impact on car use: evidence from the United States, Great Britain and Australia. Trans Res Part D: Tran Env 31:13–20
29. García-Palomares JC, Gutiérrez J, Latorre M (2012) Optimizing the location of stations in bike-sharing programs: a GIS Approach. Appl Geogr 35(1–2):235–246
30. Morizon: Rowerowy Ranking Miast (eng. Bicycle Ranking of Cities). https://www.morizon.pl/blog/rowerowy-ranking-miast-2018/. Access 15 August 2019
31. LARQ: Revolution Bike Sharing. Conference materials from Spotkanie Regionów 2018 (eng. Meeting of Regions 2018) (2018)
32. LDA Consulting: Capital bikeshare 2011 member survey report. Washington (2012)
33. Fuller D, Gauvin L, Kestens Y, Daniel M, Fournier M, Morency P, Drouin L (2011) Use of a new public bicycle share program in Montreal Canada. Amer J Prevent Med 41(1):80–83
34. Yang T, Haixiao P, Qing S (2010) Bike-sharing systems in Beijing, Shanghai and hangzhou and their impact on travel behaviour. Transp Res Board Ann Meeting, Washington, DC
35. Macioszek E, Kurek A (2020) P&R parking and bike-sharing system as solutions supporting transport accessibility of the city. Transp Probl 15:275–286
36. Nagy S, Csiszár CS (2020) Analysis of ride-sharing based on Newton's Gravity model. In: IEEE smart cities symposium, pp 1–6. IEEE Press, Prague
37. Runge A (2012) Metodologiczne problemy badania miast średnich w Polsce. Prace Geograficzne 129:83–101

Chapter 7
Factors Affecting the Choice of Transportation for School Trips

Firas Alrawi and Faisal A. Mohammed

7.1 Introduction

The growth of cities led to a change in living style, and preferences have enabled people to work in one place and live in another. One of the essential groups within urban society to be primarily affected by this direction is school students [1]. Many studies refer to the physical movement, like walking and cycling to school in the neighborhood or community as an essential part of school life [2]. In most cities worldwide, the proportion of non-motorized students' trips to school decreased during the last years, affected by different urban changes [3–5].

The measure of the city plan success for residents is by providing suitable infrastructure that contributes to reducing non-motorized transportation means and increasing reliance on walking and cycling as alternatives that promote physical activity [6].

Nowadays, it became more complicated for many people to let their children walk to school, so they became more dependent on their parents to pick them up. Therefore, that was significantly affecting their school activity, such as school attendance, punctuality, classroom activity, and adaptability in other hobbies after school [7].

F. Alrawi (✉) · F. A. Mohammed
Urban and Regional Planning Centre for Postgraduate Studies, University of Baghdad, Baghdad, Iraq
e-mail: dr.firas@uobaghdad.edu.iq

F. A. Mohammed
e-mail: faissal.abd1200b@iurp.uobaghdad.edu.iq

E. Macioszek and G. Sierpiński (eds.), *Research Methods in Modern Urban Transportation Systems and Networks*, Lecture Notes in Networks and Systems 207, https://doi.org/10.1007/978-3-030-71708-7_7

7.1.1 School Transportation

School trips, like other trips, are made by individuals through different means. For instance, a study of German cities found that nearly 55% of students using public transport for school trips, 8% using private cars or motorcycles, 17% using bicycles, and 20% walking to school. The study refers that the rates of previous transportation use differ between the winter and summer seasons, as the rates of public transport use increase in the winter while walking to school and using bicycles increases in summer with the weather enhancement [8]. Another problem is dedicated using a electric cars during this trips [9, 10].

The stage and age at which students are also influencing the choice of transportation to the school. In a suburban community in the Riverside County study, primary school students were the least likely to walk and bike to school at about 5.4% [11].

7.1.2 Transportation Selection for School Trips

Several significant factors affect the dependence of students' mobility type. Some factors are related to the student age, sex, socio-economic condition, and family structure. Other factors are extraneous impacts, mainstream transportation technology, urban infrastructure, and transportation systems applications [12].

One of the United States studies indicated that a nearby suitable school, with the availability of other safety and security factors, contributes significantly to increasing the percentage of walking to school [8].

The above indicates that parents' anxiety about traffic accidents and strangers' presence on the road are among the fundamental barriers to choosing walking as a means of transportation to school for their children [3].

In the United States, car accidents are the primary cause of death for children of school age from 5 to 16 years old, as the trip from home to school is the first cause of children's exposure to accidents [8].

The factors related to the socio-economic condition are not always a fair reflection of choosing the means of school trips. In Australia, a study showed a decrease of 80% of students who walk or cycle to school. This decrease was 77% in low living conditions regions compared to about 50% of the students at a high level of living conditions. It seems that most parents with high incomes are more aware of the importance of walking or cycling for their children as more active ways of school transportation [13]. Factors such as crime rate and insecurity in low-income neighborhoods can be associated with reduced walking to school [3].

7.2 Materials and Methods

7.2.1 The Basic Structure of the Educational System in Iraq

The pre-university educational system in Iraq consists of three stages (primary, middle, and high schools). The primary schools, which are also called elementary schools, represents the first level of education and the basic educational system. The children can join the primary school after completing sixth years old, and they spend at least six years in this stage. Primary schools are followed by middle or intermediate schools, which extends for three years as a minimum to prepares the student for the next level of secondary or high schools. High schools also extend for no less than three years until 18 years old, consisting of different majors; scientific, literary, agricultural, commercial, and industrial to prepare the students for the university stage.

7.2.2 Baghdad City as a Case Study

Baghdad's city consists of fifteen sectors, each sector consisting of several residential communities, which in turn include three to five residential neighborhoods. Baghdad is dominated by a horizontal extension with separate residential units and a few residential complexes with multi-story buildings, making it challenging to distribute services equally to all residents, including the distribution of educational services (schools) [14].

Typically, the educational services are distributed within those residential communities and neighborhoods by a primary school for each residential neighborhood, and a middle and high school within each community.

7.2.3 Data and Statistical Analysis

For data collection, a sample questionnaire was formulated to discover the significant variables affecting transportation mode choice for school trips in Baghdad. This form was directed to families that include children of school age to determine these families' behavior in selecting school transportation modes and the fundamental social and economic factors affecting determining those modes. The total number of distributed forms was about 200, covering a wide and varied city population segment. Although the sample is not large enough, it is considered sufficient due to the city population's relatively similar characteristics.

The multiple regression model was applied to the data for the three stages of schools in Baghdad to determining the essential factors of choosing transportation used in school trips. The Statistical Analysis System—SAS (2012) package was used to detect the effect of variance factors in the study parameter. The estimate of regression (Simple and Multiple Linear Regression and Prediction equations). The estimate of Simple and Multiple correlation coefficient between variables (0.05 and 0.01 probability) in this study [15]. The percentages of students in the sample were distributed to 61% of primary school students, 67% males, 33% females, 24% middle school students, 57% males, 43% females, and 15% of the high school students, 71% males 29% females. A set of questions were directed to the parents of the sample, including the essential variables and factors that may be the reason for choosing the preferred mode of transportation for school trips:

X1 Average income of the household.
X2 Car ownership per household.
X3 Number of persons per household, or household size.
X4 Average distance to school.
X5 The lack of walkways to the school.
X6 The lack of bicycle lanes to the school.
X7 Vehicle interference with the pedestrian during the school trip.
X8 Traffic density in the neighborhood during the school trip.
X9 Vehicles parked on the roadside, which impedes pedestrian movement during school trips.
X10 Crossing major streets on the school trips.
X11 Uncontrolled security situation.
X12 Road conditions.
X13 Heavy traffic around schools.
X14 Environmental conditions.

7.3 Results and Discussion

Tables 7.1, 7.2 and 7.3 show simple and multiple linear regression (Prediction equations) & correlation coefficient of (Y) on X1–X14 for primary, middle, and high schools. The null hypothesis and alternative hypothesis were examined statistically by extracting the probability value (P-value) of the factors. When (P-value) is more than the level of confidence 5%, the null hypothesis will be accepted, while when it be less or equal to 5%, the null hypothesis will be rejected and accepting the alternative hypothesis. The null hypothesis indicates that there is no relationship between variables. As for the alternative hypothesis, it indicates that there is a significant relationship between the variables, which means substantial variation and differences in the studied phenomenon.

Table 7.1 Simple and multiple linear regression (Prediction equations) & correlation coefficient of Y on X1–X14//Primary school, where *Y* represents transportation modes

Variables	Regression coefficient-b	Prediction equation	Level of sig.	R2	Correlation coefficient with Y	P-value
X1	−2.146	Y^ = 2.55 − 2.146X1	NS	0.12	−0.04	0.773 NS
X2	0.0108	Y^ = 2.17 + 0.0108X2	NS	0.08	0.02	0.928 NS
X3	−0.0426	Y^ = 2.40 − 0.0426X3	NS	0.08	0.03	0.830 NS
X4	0.0016	Y^ = 1.68 + 0.0016X4	*	0.15	0.21	0.0388*
X5	−0.0227	Y^ = 4.14 − 0.0227X5	NS	0.14	−0.04	0.790 NS
X6	−0.0109	Y^ = 3.09 − 0.0109X6	NS	0.09	0.07	0.677 NS
X7	−0.0048	Y^ = 2.62 − 0.0048X7	NS	0.11	0.06	0.682 NS
X8	−0.0199	Y^ = 3.71 − 0.0199X8	NS	0.13	0.03	0.998 NS
X9	−0.0210	Y^ = 3.99 − 0.0210X9	NS	0.10	0.04	0.802 NS
X10	0.0069	Y^ = 1.69 + 0.0069X10	*	0.1-7	0.19	0.0441*
X11	0.0073	Y^ = 1.69 + 0.0073X11	NS	0.11	0.11	0.516 NS
X12	−0.0214	Y^ = 3.93 − 0.0214X12	*	0.15	−0.04	0.786 NS
X13	−0.0069	Y^ = 2.68 − 0.0069X13	NS	0.05	−0.12	0.483 NS
X14	−0.0375	Y^ = 5.58 − 0.0375X14	NS	0.13	0.03	0.846 NS
Multiple regression						
X1–X14	Y^ = 9.90 − 2.146X1 + 0.0108X2 − 0.0426X3 + 0.0016X4-2.04X5 − 1.74X6 + 0.780X7 + 0.753X8 + 2.48X9 − 0.829X10 + 0.839X11 − O.319X12 − 0.00052X13 − 0.00024X14:*			R2 = 0.17	0.18	0.052*

*($P \leq 0.05$), NS: Non-significant

Table 7.2 Simple and multiple linear regression (Prediction equations) & correlation coefficient of Y on X1–X14//-Middle school, where *Y* represents transportation modes

Variables	Regression coefficient-b	Prediction equation	Level of sig.	R2	Correlation coefficient with Y	P-value
X1	−2.229	Y^ = 2.23 − 2.229X1	NS	0.03	−0.12	0.545 NS
X2	0.028	Y^ = 1.71 + 0.028X2	NS	0.02	−0.06	0.782 NS
X3	−0.201	Y^ = 2.84 − 0.201X3	*	0.08	−0.28	0.051*
X4	0.009	Y^ = -1.35 + 0.009X4	NS	0.07	0.16	0.397 NS
X5	0.0115	Y^ = 0.818 + 0.0115X5	NS	0.05	0.11	0.561 NS
X6	0.0079	Y^ = 1.19 + 0.0079X6	NS	0.02	0.10	0.613 NS
X7	0.0064	Y^ = 1.17 + 0.0064X7	NS	0.03	0.09	0.629 NS
X8	−0.0237	Y^ = 3.68 − 0.0237X8	*	0.09	−0.24	0.0487*
X9	0.0178	Y^ = 0.433 + 0.0178X9	NS	0.05	0.14	0.491 NS
X10	0.0089	Y^ = 1.17 + 0.0089X10	NS	0.02	0.13	0.497 NS
X11	−0.0012	Y^ = 1.82 − 0.0012X11	NS	0.02	−0.02	0.929 NS
X12	0.0053	Y^ = 1.33 + 0.0053X12	NS	0.07	0.06	0.876 NS
X13	0.006	Y^ = 1.31 + 0.006X13	NS	0.05	0.03	0.876 NS
X14	0.0347	Y^ = 4.89 + 0.0347X14	*	0.07	0.29	0.0446*
Multiple regression						
X1–X14	Y^ = 11.21 − 2.229X1 + 0.028X2 − 0.201X3 + 0.009X4 − 0.020X5 + 0.008X6 − 0.090X7 − 0.092X8 + 0.081X9 − 0.037X10 − 0.0022X11 − 0.00035X12 + 0.004X13 + 0.007X14:*			$R^2 = 0.34$	−0.23	0.0497 *

* ($P \leq 0.05$), NS: Non-significant

Table 7.3 Simple and multiple linear regression (Prediction equations) & correlation coefficient of Y on X1–X14//High school, where *Y* represents transportation modes

Variables	Regression coefficient-b	Prediction equation	Level of sig.	R2	Correlation coefficient with Y	P-value
X1	−0.000011	Y^ = 28.40 − 0.000011X1	*	0.23	−0.48	0.027*
X2	0.000024	Y^ = 1.65 + 0.000024X2	**	0.99	0.99	0.0001**
X3	−8.12	Y^ = 51.59 − 8.12X3	*	0.22	−0.47	0.031*
X4	−0.139	Y^ = 57.94 -0.139X4	**	0.56	−0.75	0.0001**
X5	0.252	Y^ = −17.86 + 0.252X5	**	0.95	0.97	0.0001**
X6	0.613	Y^ = −52.52 + 0.613X6	NS	0.03	0.16	0.475 NS
X7	0.313	Y^ = −23.86 + 0.313X7	NS	0.03	0.11	0.608 NS
X8	0.104	Y^ = −2.93 + 0.104X8	NS	0.02	0.12	0.609 NS
X9	−0.737	Y^ = 53.57 − 0.737X9	**	0.49	−0.70	0.0004**
X10	−0.263	Y^ = 24.82 − 0.263X10	*	0.20	−0.45	0.042*
X11	−0.379	Y^ = 29.66 − 0.379X11	*	0.24	−0.49	0.0234*
X12	−0.311	Y^ = 27.08 − 0.311X12	*	0.22	−0.46	0.0321*
X13	−0.440	Y^ = 40.01 − 0.440X13	**	0.37	−0.60	0.0038**
X14	0.156	Y^ = −8.17 + 0.156X14	NS	0.08	0.12	0.608 NS
Multiple regression						
X1–X14	Y^ = 6.92 − 0.000011X1 + 0.000024X2-8.12X3 − 0.139X4 + 0.010X5 − 0.435X6 + 1.34X7 + 0.007X8 − 0.771X9-0.385X10 − 0.174X11 − 0.0012X12 − 0.0002X13 + 0.0073X14:**			R2 = 0.91	0.54	0.0089**

*($P \leq 0.05$), **($P \leq 0.01$), NS: Non-significant

7.3.1 Primary Schools

After applying the multiple linear regression model to primary school students' data, a set of factors that determine their transportation mode to the school were extracted, Table 7.1.

It turns out that two main factors influence the determination of school transportation means for this group of students.

The first factor is X4, the distance from home to school, which correlates positively with mechanical means as an option for transportation to school instead of walking with a probability value of 0.0388. The regression coefficient reached 0.0016, which is expected as the reliance on vehicle transportation increases, the more distant the school is. However, the strange thing is that only 30% of the study sample of those within the range of 500 m (Standard set distance for elementary school) who walk to school.

The second factor was represented by the presence of major streets on the road to schools X10, the probability value of which was 0.0441, and the regression coefficient was 0.0069, which shows a positive correlation. It indicates that any extra major streets on the way to school, representing a danger to students of this age group, increase vehicle users' percentage on school trips.

The effect of other factors is much less than the factors above, and the regression Eq. 7.1 can be used to determine the type of future means of transportation for primary school students.

$$\begin{aligned}\hat{y} = 9.90 - 2.146x1 + 0.0108x2 - 0.0426x3 + 0.0016x4 - 2.04x5 - 1.74x6 \\ + 0.780x7 + 0.753x8 + 2.48x9 - 0.829x10 + 0.00144x11 \\ + 0.0839x12 - 0.319x13 - 0.00052x14 - 0.00024x15 \qquad (7.1)\end{aligned}$$

7.3.2 Middle Schools

For middle school students, the regression analysis results showed three main factors related to the characteristics of those students, which affect their choice of school transportation, as shown in Table 7.2.

The number of family members X3 was the first factor, the probability value of which was 0.051, and the regression coefficient was −0.201, which means that there is a negative correlation for this factor with reliance on mechanical transport. In other words, the increase in family members contributes to increasing dependence on walking to school. The second factor, X8, represented by severe crowding, which has a probability value of 0.0487, and a regression coefficient with a negative correlation of −0.0237. It is believed that the negative correlation is because students at this age can walk even with the presence of heavy traffic in the neighborhood to avoid delaying the trip time, especially for those who live near the school. For the third factor related to weather conditions, X14 showed a positive correlation with a regression coefficient of 0.0347 and a probability value of 0.0446.

The other factors had less effect than the three mentioned factors, which appear in Eq. 7.2.

$$\begin{aligned}\hat{y} = 11.21 - 2.229x1 + 0.028x2 - 0.201x3 + 0.009x4 - 0.020x5 + 0.008x6 \\ - 0.090x7 - 0.092x8 + 0.081x9 - 0.037x10 + 0.068x11 \\ - 0.0022x12 - 0.00035x13 + 0.004x14 - 0.007x15\end{aligned} \tag{7.2}$$

7.3.3 *High Schools*

Table 7.3 shows the regression analysis results of high school students' characteristics with their preference for transportation to school.

While the results showed a negative correlation with factors related to income rate, family size, distance to school, roadside parked vehicles, crossing major roads, security status, road conditions, and traffic density. In other words, it seems that the age of this group of students gives them more possibility to walk to school despite some of the difficulties mentioned, as the results are shown in Table 7.3. The regression Eq. 7.3 can be used to determine future means of transportation for high school students.

$$\begin{aligned}\hat{y} = 6.92 - 0.000011X1 + 0.000024X2 - 8.12X3 - 0.139X4 + 0.010X5 \\ - 0.435X6 + 1.34X7 + 0.007X8 - 0.771X9 - 0.385X10 \\ - 0.020X11 - 0.174X12 - 0.0012X13 - 0.0002X14 + 0.0073X15\end{aligned} \tag{7.3}$$

7.4 Conclusion

By applying the multiple linear regression model, this chapter clarified the extent of the relationship between school students in Baghdad in their three levels (primary, middle, and high schools), and the influencing factors for determining transportation means for school trips. The school stage difference indicated a difference in the influence of social and economic factors on determining the means of transportation to school. The increase in the distance from the primary school and the presence of obstacles such as the major streets had the most considerable effect, among other factors, on increasing reliance on automated transportation for elementary school students.

While factors such as the increase in the family size and traffic density in the residential neighborhood positively affected walking as an alternative to transportation on middle school trips. The high schools, factors such as vehicle ownership, and lack of infrastructure associated with walking were among the most prominent factors that lead to increased dependence on the vehicle on high school trips.

To motivate students and their parents and restore confidence in a way such as walking and cycling for school trips. Several measures should be taken regarding the development of the residential community's infrastructure by taking care of pedestrian paths and planting them, establishing bicycle lanes, and separating the movement of vehicles from pedestrians. Also, providing school buses to students who cannot walk to school due to distance or other reasons. All of this will encourage an active journey (walking or cycling) to school within standard distances and reduce the use of mechanical means.

Acknowledgements To all the staff of the Institute of International Education (IIE), your support has been outstanding.

References

1. Aasmoah GA, Okyere SA, Senayah EAK (2015) Factors influencing school travel mode choice in Kumasi Ghana. Int J Dev Sustain 4(1):01–17
2. Garrard J (2016) Walking, riding, or driving to school: what influences parents' decision making? Technical report, South Australian Department of Planning, Transport, and Infrastructure
3. Khalil NS (2013) Factors affecting students walking to school: case study of two middle schools in Lincoln, Nebraska. A dissertation submitted for the degree of Master of Community and Regional Planning, University of Nebraska, Lincoln, Nebraska
4. Chebanyuk K, Prasolenko O, Burko D, Galkin A, Lobashov O, Shevchenko A, Usami DS, Persia L (2020) Pedestrians influence on the traffic flow parameters and road safety indicators at the pedestrian crossing. Transp Res Procedia 45:858–865
5. Naumov V, Samchuk G (2017) Class library for simulations of passenger transfer nodes as elements of the public transport system. Procedia Eng 187:77–81
6. Stark J, Bartana IB, Fritz A, Unbehaun W, Hössinger R (2018) The influence of external factors on children's travel mode: a comparison of school trips and non-school trips. J Transp Geogr 68:55–66
7. Shokoohi R, Hanif NR, Dali M (2012) Influence of the socio-economic factors on children's school travel. Procedia Soc Behav Sci 50:135–147
8. Al-Alwan AM (2013) The impact of the availability of safety and security factors in school transportation. J King Abdulaziz Univ Environ Des 7:03–50
9. Macioszek E, Sierpiński G, Staniek M (2017) Analysis of trends in development of freight transport logistics using the example of Silesian Province (Poland)—a case study. Transp Res Procedia 27:388–395
10. Sierpiński G, Macioszek E (2020) Equalising the levels of electromobility implementation in cities. In: Mikulski J (ed) Research and the future of telematics, vol 1289. Communications in computer and information science. Springer, Heidelberg, pp 165–176
11. McMillan TE (2003) Walking and urban form: modeling and testing parental decisions about children's travel. A dissertation submitted for the degree of doctor of philosophy in urban and regional planning, University of California, Irvine, The USA
12. Alrawi F (2017) The importance of intelligent transport systems in the preservation of the environment and reduction of harmful gases. Transp Res Procedia 24:197–203

13. Conlon F (2013) Getting to school factors affecting choice of active travel modes in the trip to school. A dissertation submitted for the degree of Master of Public Health, University of Otago, Wellington, New Zealand
14. Alrawi F (2018) Measuring the relative importance of applying engineering solutions to urban traffic intersections: a planning perspective. Scientific Journal of Silesian University of Technology. Series Transport, 100, pp 05–13
15. SAS (2012) Statistical Analysis System, User's Guide. Statistical, Version 9.1th ed. SAS. Inst. Inc. Cary. N.C. USA

Chapter 8
Research on Parents' Attitude Towards Children Safe Transportation: The Cross-Sectional Survey Method

Iryna Tkachenko, Andrii Galkin, Davide Shingo Usami, Veronica Sgarra, and Luca Persia

8.1 Introduction

Compared to previous years, fewer people lost their lives on EU roads in 2019, according to latest figures published by the European Commission [1]. An estimated 22,800 people died in a road crash last year, almost 7000 fewer fatalities than in 2010—a decrease of 23%. Compared to 2018, the number fell by 2%. With an average of 51 road deaths per 1 million inhabitants, Europe remains by far the safest region in the world when it comes to road safety. Safe routes and vehicles are one of the ways to reduce the number of road accidents victims [2–5]. According to [6] the same dynamics we can see regarding to child injuries: a total of 530 children (aged 15 years and below) were killed in motor vehicle accidents in 2018 (2.12% of the total number of fatalities), 593 children—in 2017 (2.32%), 626 children—in 2016 (2.46%) accordingly. However, each year more than 100,000 children under the age of 15 are killed on European roads and are injured. The worst performing countries in children motor vehicle injuries are France, Germany, Romania and Poland [7].

According to Directive 2003/20/EC [8], seat belts must be used in all vehicles. Children over 1.35 m can use an adult seat belt. Those under 1.35 m must use equipment appropriate to their size and weight when travelling in cars or lorries. It is

I. Tkachenko (✉) · A. Galkin
O. M. Beketov National University of Urban Economy in Kharkiv, Kharkiv, Ukraine

D. S. Usami · V. Sgarra · L. Persia
Research Centre for Transport and Logistics Sapienza University of Rome, Rome, Italy
e-mail: davideshingo.usami@uniroma1.it

V. Sgarra
e-mail: veronica.sgarra@uniroma1.it

L. Persia
e-mail: luca.persia@uniroma1.it

E. Macioszek and G. Sierpiński (eds.), *Research Methods in Modern Urban Transportation Systems and Networks*, Lecture Notes in Networks and Systems 207, https://doi.org/10.1007/978-3-030-71708-7_8

now against the law to use a rear-facing child seat on the front passenger seat–unless the airbag has been deactivated.

The use of child restraint systems (CRSs) is one of the most effective ways to prevent death and severe injuries among children in case of road accident. The appropriate and correct use of CRS can reduce the risk of death and serious injuries in toddlers and infants from 71 to 54% [9]. Children aged 4–8 years buckling into a CRS seem to be 45% less likely to get injured than children of similar ages using a vehicle seat belt alone [10].

Globally [11], it is more largely accepted not to wear a seat belt than not to use an appropriate CRS. On a regional level, Europe presents the lowest level of self-declared acceptance for driving without a seat belt (average of 4%). On a national level, the acceptance rates in the different European countries range from 1.3% in Ireland to 8.4% in Poland. North America is the region with the lowest acceptance for not using an appropriate CRS, with an average of 1% acceptance rate. The highest acceptance for both restraint devices is found within the African region with on national level Egypt revealing the highest acceptance rates: 25.5% for failure to wear a seat belt and 21.8% for transporting children without a CRS. According to an international survey, 15% of respondents who transported a child in 20 European countries declared not to have used a CRS at least once in the previous 30 days.

Despite the introduction of amendments to the Ukrainian's Road Traffic Law, which went into effect in 2019 and mandate that it is forbidden to transport children under 12 and height below 145 cm without a safety car seat, the prevalence of CRS use is very low.

According to official statistics of Ukrainian Patrol Police [12]: 4656 road accidents involving children (2.8% of the total number of accidents) occurred in 2019, among them more than 4435 children were injured and 164 died (4.8% of the total number of crash fatalities). The rate of children injured in road accidents in Ukraine is significantly higher than in Europe, 176 children were killed in 2018 (5.3% of the total number of crash fatalities), 175 children were killed in 2017 (5.0% of the total number of crash fatalities). Children under 7 years of age who have not been properly restrained have the highest number of fatal injuries in road accidents.

One of the most common methods for studying the determinants affecting awareness, perception and usage of CRS is cross-sectional study involves looking at data from a population at one specific point in time and allows researchers to look at numerous characteristics at once (age, income, gender, etc.). This method is often used to make inferences about possible relationships or to gather preliminary data to support further research and experimentation. It has such benefits as: possibility of relatively quick conduction, multiple outcomes can be researched at once, suitable for descriptive analysis; prevalence for all factors can be measured etc.

The aim of this study was to describe the importance of application the cross-sectional analysis method to research different variables that influences child passengers sitting behavior and CRS using.

8.2 Literature Review

The method of cross-sectional surveys is a type of observational study design. It is commonly used for population-based surveys to make inferences about possible relationships or to gather preliminary data to support further research in the different field of science and is useful to designing a cohort study [13].

This method as opposite to cohort study, where participants are selected based on the exposure status of the individual and then followed over time to evaluate for the occurrence of the outcome of interest and case-control study, where participants are selected based on their outcome status, allows to investigate the outcome and the exposures in the study participants at the same time.

Within the research on the use of CRS, cross sectional studies have been adopted for multiple purposes. Typical research questions investigated are:

- how common is e.g. the use of the system in a population, or its appropriate use, or a specific position of the system in the car, or the intention to use the system in the next future (aimed at mother in pregnancy)?
- what factors are associated with these behaviors/beliefs/conditions, e.g. factors influencing the use/not use of the system, or its appropriate use, or the position of the system in the car, or the intention to use the system?

Implementation of this method has become widespread not only in Europe, but throughout the world. A large body of work has characterized child restraint use in Australia using population-based representative telephone surveys [14], school-based surveys or direct observations [15].

Analysis methods like logistic regression, multivariate regression, chi-square test provide an opportunity not only to establish the relationship between parents'/careers' social-demographic characteristics and their attitude and perception of child safety transportation but to calculate it as a measure of association in terms of Odds Ratio, describe data and explain the relationship between variables.

Examples of applications of these statistical methods for cross sectional data analysis are reported in the Table 8.1.

The Theory of Planned Behaviour (TPB) from Azjen is frequently adopted to model the intention to use CRS.

For instance, using this framework and multivariate analysis allowed to identify the determinants of CRS use that have mostly dealt with individual characteristics (i.e. age, gender, height and weight of the child), travel characteristics (i.e. driving distance, night/day travel, driver's use of a car seat belt, number of children traveling in the car), parents' attitudes, perceptions and obstacles to use, and child car restraint legislation [19].

Focusing on factors affecting the use of CRS through cross sectional studies, it was found that the main reasons for the non-use of CRSs are due to the following factors:

- lack of use of seat belts by parents,
- toddler ages,

Table 8.1 Cross sectional data analysis methods

Method	Application	References
Multivariate logistic regression	Binary multiple logistic regression was used to establish interlinkages between parents' unsafe behaviour (not using seat belts for themselves and not using CRS for children) as dependent variable and social-demographic and motivation factors as independent variables [17] Multinomial logistic regression describes the independent variable as a discrete one that can take a finite number of values, more than two. It was used for example to study the association of demographic characteristics, local CRS legislation, knowledge of CRS laws, and use of booster seats with membership to three groups of parents labeled as Benefit Sensitive, Context Sensitive and High Risk [19]	[14, 15, 17, 19, 21, 22]
Hi-square tests or Fisher's tests	Used to check if there is any difference in current behaviours among subjects of different age groups [17]. Fisher's test is used if the expected number of observations in a cell <5	[17, 19]
Structural equation model (SEM)	SEM measures the relationship among contribution of different observable variable as well as latent variable. It was used to identify the most significant predictors of the intent to use a booster seat and reported use of booster seats [18]	[18]
Discrete choice conjoint analysis	In choice-based conjoint surveys, informants make choices among a controlled set of potential products or services with the aim to determine which combination of attributes is most influential on respondent decision making In [19] this approach supported identifying factors influencing the decision of parents to use CRSs	[19]
Hierarchical multinomial logistic regression	On the first level each parent's variance was tested: individual characteristics and parent-child relationship characteristics. On the second level, under the assumption of dependence between parents (influenced by residence in the same neighborhood and its characteristics) the variance between neighborhoods was tested [23]	[23]

- parents education,
- lack of knowledge rules,
- economic availability,
- vehicle crowding,
- lack of risk perception.

Collarile et al. [17] evaluated the prevalence of the correct use of CRS, in correlation with the level of parental information and parental driving behaviors in a population of North-Eastern Italy. A multivariable logistic regression was used to identify factors influencing correct restraint use. Results showed that the frequency of habitual use decreases with increasing child's age, and that increased child's age was associated with decreased likelihood of correct restraint use. Moreover, the correct restraint use and parental information on child restraints, and current parental seat belt use were positively associated with correct use.

In 2017, Kanat et al. [23] have analysed individual, parent-child relationships, and neighborhood characteristics through comprehensive approach in which data were subjected to multilevel modeling. Findings demonstrated that age-appropriate car restraint use for children aged 4–7 years old remains the lowest among all age groups and that the determinants positively related to CRS use were individual and parental: the number of children in the family, the parents' car seat belt use, parental knowledge of children's car safety principles, and a highly authoritative parenting style.

An analysis of a CRS use in low socio-economic areas of urban Sydney was presented in 2013 by Keay et al. [14]. The study has investigated factors associated with restraint use to inform the development of education and distribution programs to support new Australian legislation on child passengers among families in low socio-economic areas of metropolitan Sydney. Predictors of self-reported appropriate use were explored using logistic regression. Findings showed that inappropriate use among children in this age group was more likely when the career was <36 years, and in families with $\geq$3 children. Non-use was associated with low household income, in families with $\geq$3 children and families where a language other than English is spoken at home.

Agran et al. [24] in 2006, have explored hassles that are associated with child safety seat non-use, focusing on groups with violators of the California Child Passenger Safety Law provided data to construct the 29-item Hassles Scale. The scale was used in an interview that was conducted with 132 parents who were cited for violation of the law and whose children were 12–47 months of age and weighed 20–40 lb. The sample was 86% Latino, 45% Spanish-speaking, and 55% with income < dollar 30,000. The 31% of the parents reported that the child did not always use a child safety seat because of crowding/inconvenience (e.g. child safety seat takes up too much room, too many passengers) and for the child's behaviour (e.g. resists, gets out of seat).

Systematic reviews [19] have concluded that educational interventions, combined with benefits or the distribution of free booster seats, lead to gradual increases in using of CRS, especially for 4–9-year-old children.

Concerning the lack of risk perception, in particular in terms of misperception of children's physical resistance, Ebel et al. [25] in 2006 has conducted a study with the objective to determine barriers and facilitators for booster seat use among Latino families. The research consisted in performing one-to-one elicitation interviews with 56 mothers and 35 fathers of booster-eligible Latino children in an urban county and a rural county in Washington State. Findings have shown how half of the parents did not consistently use booster seats for child occupants between 4 and 8 years of age. Facilitators for booster use included affordability, ease of use, and children liking the seat. Barriers were the belief that the child was too big/old, perceived child resistance, and cost.

8.3 Methods

8.3.1 Study Design and Participants

The data used in this study were obtained through cross-sectional observational surveys conducted in Kharkiv, Ukraine in October 2019. Two types of observation locations were selected to represent children aged 3–10 years: kindergarten, attended by children aged 3–6 years; primary school, attended by children aged 7–10 years. The study locations were selected with a consideration to cover city areas of different economic status (e.g. middle-income and high income).

8.3.2 Data Collection

Two trained interviewers provided brief verbal explanations about the objectives of the study and how to complete a questionnaire and collected them when completed to parents during parent-teacher meetings in October 2019. Then an email survey consisting of 40 questions concerning knowledge, attitudes and reasons for appropriate/inappropriate using or not using a CRS was conducted among parents, who usually drop off or pick up their children. This questionnaire included different characteristics of parents and their children.

Individual characteristics included parental knowledge regarding child's safety in the car, driving behavior, socio-demographic and driving characteristics.

Socio-demographic variables included child's age, parents' age and gender, the level of education and income, employment status and the number of children in the car that must be fastened.

Driving variables included presence of driving license, using a seatbelt while driving, average daily distance and trips frequency, violation of traffic rules and attitude towards increasing fines for it.

Declared behaviour of parents regarding their child's safety in the car was assessed by questions such as "In the last year, how often did your child use a CRS?" or "In the last year, how often did your child use a seatbelt?" (with an answers: never, rarely, sometimes, almost always, always).

Driving behavior was assessed by questions such as "Is it possible that drivers may exceed the set speed limit by more than 20 km/h in residential or pedestrian areas of your city?", "Is it possible in your city that drivers can cross the intersection at a forbidden traffic light?" (with an answers: possible or impossible) or "In the last year, how often did you talk on mobile phone while driving?", "In the last year, how often did you use a seatbelt while driving?" (with an answers: never, rarely, sometimes, almost always, always).

8.4 Results

The final sample of 100 parents' questionnaires was included to observation: 87% of respondents were parents, 13%—relatives or hired professional drivers. More than two thirds of respondents (64%) were male drivers aged 30–55 years and one thirds of respondents (36%) were female drivers aged 23–50 years. Most of them (72%) reported that observance of traffic rules is not properly monitored by the police but at the same time they decided that traffic fines are too harsh (67%). Statistical distribution of parental attitudes towards seatbelt and CRS using is shown in Figs. 8.1 and 8.2.

Young drivers (OR 4.216, 95% CI = 1.799, 9.898) with a low-income level (OR 2.794, 95% CI = 1.230, 6.346) whom daily travel distance is up to 10 km (OR 5.435, 95% CI = 2.020, 14.62) were less likely to buckle their children to CRS. Parents

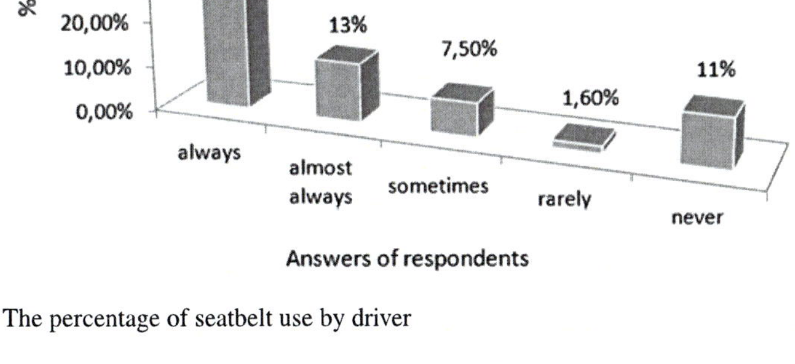

Fig. 8.1 The percentage of seatbelt use by driver

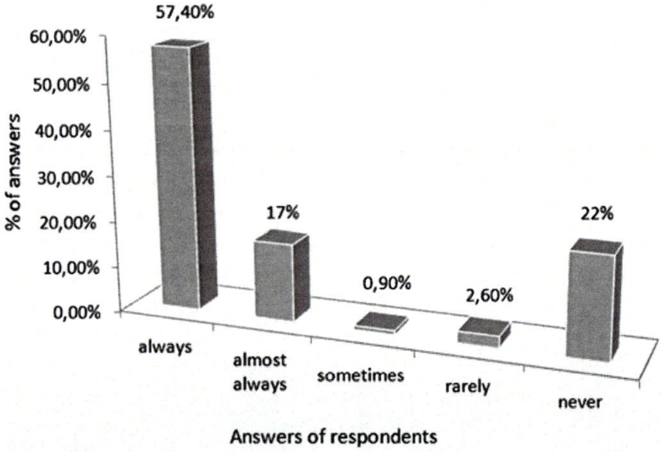

Fig. 8.2 The percentage of CRS use by children

with high education level who have only one child into the car (OR 3.884, 95% CI = 1.590, 9.488) under 5 years old (OR 12.745, CI = 0.534, 36.27) were more likely to use CRS (Table 8.2).

The various recorded characteristics associated with driver and child passenger are summarized in a percentage distribution in Table 8.2.

This study also found that CRS use was significantly influenced by the driver's belt use and that men were more likely to fixed children properly.

The second parent-teacher meeting dedicated to improving the safety of the child during transportation was held after processing of the questionnaire results and identifying the main factors affecting parents' degree of responsibility for the use of CRS. The reasons of not using CRS among children with the age from 3 to 10 years were identified in personal conversation with parents, the main which are: high price (69.7%); lack of awareness (26.7%); and lack of a system of fines (15.9%). Some parents had their younger children on their laps, perhaps wrongly believing that holding children on an adult's lap is a better way to protect them.

Interactive games were conducted with children in order to disseminate, motivate and increase child's level of responsibility for appropriate using of CRS.

Two months after meetings, the second survey with the same questions was conducted, which showed significant increasing of parental' perceptions and attitudes to the issues of road safety according to which the use of CRS increased from 57.4 to 74% and the use of seat belt by drivers increased from 66.7 to 74.2% of total respondents.

As a result of the evaluation and positive dynamics, the management of the educational institutions has taken decision to organize annual educational programs in the first week of study year for parents and their children to promote the use of CRS.

Table 8.2 Individual characteristics affecting on CRS using

		OR (95% CI)	
Indicators	The total number of respondents	Not using or inappropriate using of CRS	Using of CRS
Gender: – Male – Female	64 36	0.636 (0.253, 1.604)	1.571 (0.624, 3.960)
Age of driver: – 18–34 – 35–50	42 58	4.216 (1.799, 9.898)	0.237 (0.101, 0.556)
Education: – Secondary – High	5 95	0.917 (0.146, 5.743)	1.091 (0.174, 6.835)
Monthly income: – Up to 10000 UAH – >10000 UAH	45 55	2.794 (1.230, 6.346)	0.358 (0.158, 0.813)
Number of children in the car: – 1 – >2	69 31	0.257 (0.105,0.629)	3.884 (1.590, 9.488)
Age of child: – Under 5 years old – 5–12 years old	69 31	0.078 (0.028, 0.223)	12.745 (0.534, 36.27)
Daily travel distance: – Up to 10 km – >10 km	48 42	5.435 (2.020,14.62)	0.184 (0.068, 0.495)

8.5 Conclusions

This study has combined research of using a child restraint system with parental responsibilities observation. Application of the method of cross-sectional surveys to research parents' attitude towards children safe transportation has given an opportunity to establish an association between different socio-demographic characteristics of drivers and frequency and correctness of use child restraint system. Number of children in the family, their age and driver level of income were found to be the most influential demographic characteristics to support awareness and educational campaigns as well as compliance with the law regarding child car seat use.

The study provides important information about the safety regulations for child passengers, about the knowledge, attitude, interest and willingness of parents to investigate the safety benefits of CRS, as well as the need and design promotional campaigns to educate the public on the need to appropriately position and restrain child passengers.

Such information is essential for government organizations in charge of road safety and other stakeholders to support and guide policies and programs to improve road safety practices, design effective counter measures and intervention programs to minimize the high level of death and injuries among children of different age.

References

1. An official website of the European Union. https://ec.europa.eu/transport/media/news/2020-06-11-road-safety-statistics-2019_en
2. Makarova I, Pashkevich A, Shubenkova K (2018) Safe routes as one of the ways to reduce the number of road accidents victims. In: Macioszek E, Sierpiński G (eds) Recent advances in traffic engineering for transport networks and systems, vol 21. LNNS. Springer, Cham, pp 73–84
3. Tubis A, Werbińska-Wojciechowska S (2014) Safety measure issues in passenger transportation system performance: case-study. In: Steenbergen RDJM, van Gelder PHAJM, Miraglia S, Vrouwenvelder ACWM (eds) Safety, reliability and risk analysis, pp 1309–1316 (2014)
4. Sierpiński G, Macioszek E (2020) Equalising the levels of electromobility implementation in cities. In: Mikulski J (ed) Research and the future of telematics, vol 1289. Communications in computer and information science. Springer, Heidelberg, pp 165–176
5. Macioszek E, Sierpiński G (2020) Charging stations for electric vehicles—current situation in Poland. In: Mikulski J (ed) Research and the future of telematics, vol 1289. Communications in computer and information science. Springer, Heidelberg, pp 124–137
6. EuroStat. https://ec.europa.eu/eurostat/web/transport
7. Traffic Safety Basic Facts 2018-Children. https://ec.europa.eu/transport/road_safety/sites/roadsafety/files/pdf/statistics/dacota/bfs2018_children.pdf
8. Directive 2003/20/EC of the European Parliament and of the Council of 8 April 2003 amending Council Directive 91/671/EEC on the approximation of the laws of the Member States relating to compulsory use of safety belts in vehicles of less than 3,5 tones. https://eur-lex.europa.eu/eli/dir/2003
9. Garces AQ, Coimbra IB, Silva DS (2016) Transporting children in cars and the use of child safety restraint systems. Acta Ortop Bras 24:275–278
10. Arbogast KB, Jermakian JS, Kallan MJ, Durbin DR (2009) Effectiveness of belt positioning booster seats: an updated assessment. Pediatrics 124(5):1281–1286
11. E-Survey of Road Users' Attitudes. https://www.esranet.eu/en/publications/
12. Patrol Police of Ukraine. http://patrol.police.gov.ua/
13. Miller FP, Vandome AF, McBrewster J (2011) Cross-sectional study. VDM Publishing
14. Keay L, Hunter K (2013) Child restraint use in low socio-economic areas of urban Sydney during transition to new legislation. Accid Anal Prev 50:984–991
15. Koppel S, Charlton JL (2009) Child restraint system misuse and/or inappropriate use in Australia. Traffic Injury Prevent 10(3):302–307
16. Ajzen I (1991) The theory of planned behaviour. Organ Behav Hum Decis Process 50:179–211
17. Collarile P, Valent F, Di Bartolomeo S, Barbone F (2008) Changes in child safety restraint use and parental driving behaviours in Italy. Acta Pædiatric 97:1256–1260
18. Bruce BS, Snowdon AW (2013) Predicting parents' use of booster seats. Injury Prevent 2011:11–16
19. Cunningham CE, Bruce BS, Snowdon AW, Chen Y, Kolga C, Piotrowski C, Warda L, Correale H, Clark E, Barwick M (2011) Modeling improvements in booster seat use: a discrete choice conjoint experiment. Accid Anal Prev 43(6):1999–2009
20. Jeffreya J, Whelanb J, Pirouzc DM, Snowdon AW (2016) Boosting safety behaviour: descriptive norms encourage child boosterseat usage amongst low involvement parents. Accid Anal Prev 92:184–188
21. Rok SM, Korošec A, Bilban M (2017) The influence of parental education and other socio-economic factors on child car seat use. Zdr Varst 56(1):55–64

22. James LJ, Karl K (1996) Restraint use by children involved in crashes in Hawaii. Transp Res Record J Transp Res Board 1560(1):8–12
23. Kanat SS, Gofinb R (2017) An ecological model to factors associated with booster seat use: a population based study. Accid Anal Prev 108:245–250
24. Agran PF, Anderson C, Winn D (2006) Development of a child safety seat hassles scale in a largely low-income latino population. Pediatrics 118(1):85–91
25. Ebel B, Coronado GD (2006) Child passenger safety behaviors in Latino communities. J Health Care Poor Underserved 17(2):358–373

Chapter 9
Speed-Related Surrogate Measures of Road Safety Based on Floating Car Data

Jiří Ambros, Chris Jurewicz, Anna Chevalier, and Veronika Valentová

9.1 Introduction

Traditionally, road safety has been assessed based on accumulation of crashes. However, such approach has several limitations, including the reactive nature, and the statistically low occurrence of crashes. To overcome limitations of crash-based road safety analyses, researchers are looking for surrogate measures of safety (SMoS [1]). In parallel, a number of studies have recently focused on emerging sources of floating car data (FCD, also known as probe vehicle data). This data is sampled from vehicle fleets, i.e., "collected by the vehicles themselves" [2]. Several authors investigated possibilities of using FCD to extract surrogate measures for road safety studies. However, their results were not conclusive and opened up several questions related to definitions of SMoS, validity, etc. The aim of this chapter is to critically review this field and discuss challenges and opportunities related to using FCD as a source for SMoS. FCD as data "collected by the vehicles themselves" generally comes from two data sources:

J. Ambros (✉) · V. Valentová
CDV—Transport Research Centre, Brno, Czech Republic
e-mail: jiri.ambros@cdv.cz

V. Valentová
e-mail: veronika.valentova@cdv.cz

C. Jurewicz
Transport Accident Commission, Melbourne, Australia
e-mail: Chris_Jurewicz@tac.vic.gov.au

A. Chevalier
Australian Road Research Board, Sydney, Australia
e-mail: anna.chevalier@arrb.com.au

E. Macioszek and G. Sierpiński (eds.), *Research Methods in Modern Urban Transportation Systems and Networks*, Lecture Notes in Networks and Systems 207, https://doi.org/10.1007/978-3-030-71708-7_9

- navigational sensors, i.e., Global Navigation Satellite Systems (GNSS) or through mobile phone triangulation (also known as Global System for Mobile Communication [GSM] data or cellular FCD). GNSS signals may be registered by a portable device (smartphone or satellite navigation), or an in-vehicle data recorder (IVDR),
- additional sensors (e.g., accelerometer, odometer, gyroscope, magnetometer) or a connection to a Controller Area Network (CAN bus), which enables recording data from other car sensors (odometer, fuel consumption, engine performance, etc.—such enhanced data is also called extended floating car data, xFCD),
- the following review focuses on FCD as a source of measures to study driving behaviour and safety. In this regards the most frequently used indicators are speed (obtained from GNSS data) and its derivations such as acceleration and jerk (from GNSS or accelerometer data).

Note that this review does not include the FCD collected from other in-vehicle devices. While these offer numerous interesting data sources (video monitoring, recording pedal positions, surrounding road users, etc.), they are usually limited to smaller fleets or individual vehicles. On the contrary, the review focuses on FCD routinely available from large vehicle fleets fitted with navigation devices or carrying mobile phones.

In general, FCD consists of data points, characterized by their geo-location and time. Since data is collected at known sampling rate, speed may be simply calculated. Speed is interesting from the perspective of SMoS, since there is well documented relationship between speed and safety. For example, the Power Model [3–5] related the effects of mean speed changes to the number of crashes of different severity, and was validated in various road environments [6, 7]. In addition, speed may be used to further calculate derivatives such as acceleration or jerk.

Using FCD to obtain speed data has several advantages over traditional measurement techniques, such as roadside traffic counters, fixed loops or tube counters. FCD enables network-wide data collection, as well as availability of historical data, which is an ideal source for before-after studies.

The goal of this chapter is to identify opportunities and challenges regarding using speed-related and FCD-based indicators (speed, acceleration, jerk) to develop SMoS. These will be demonstrated on three example applications (use cases):

- using speed or speeding as a safety performance indicator, collected in a representative network of sites, for example to evaluate countermeasures (national speed limit changes, campaigns, enforcement, etc.) and observe long-term national safety trends,
- using speeding or harsh deceleration to identify dangerous events or assess driving behavior,
- using speed or harsh deceleration (or derived indicators) to identify and assess high-risk sites, as well as safety variations, for example investigating the impact on driving behavior of curve radii, road widths, or traffic calming measures.

Compared to previous reviews, which focused mainly on uses of FCD and GSM for traffic monitoring (e.g., [2, 6, 7]), this review focuses primarily on the safety perspective. Despite the mentioned benefits, it should be remembered that these FCD were collected to serve purposes not related to safety (such as for navigation and/or traffic monitoring). To investigate whether FCD may be confidently used in road safety research, the differences between the original purposes and mentioned research approaches and their implications will be discussed.

9.2 Methodology

This chapter undertook a review of available literature across a range of subjects and study types pertaining to FCD-based speeds. Given the exploratory nature of this review, sources with unclear methodology or FCD sources were set aside. All other manuscripts were reviewed to extract qualitative and quantitative information.

The review parameters included:

- retrieved sources:
  - papers from Web of Science, PubMed, Scopus and TRID databases, including review of references cited in reviewed manuscripts.
  - "grey literature": the Australian Road Research Board (ARRB) Knowledge Base, institute reports, naturalistic driving study (NDS)/field operational test (FOT) project deliverables.
  - proprietary data specification sources.
- keywords: floating car data, probe vehicle data, speed, safety,
- language: English,
- timeframe restriction: none.

The review findings were organized into logical topics pertaining to the collection and potential use of FCD to develop SMoS. Information related to each of these topics were synthesized to develop conclusions about opportunities and challenges. These topics were as follows:

- sampling rate,
- study size,
- free-flow speed determination,
- reliability,
- validity,
- use cases,

9.3 Results

Using the mentioned review parameters, 86 and 57 items were identified in scientific databases and in grey literature, respectively (143 in total). In the second step, the selection was screened for studies out of scope of this review, such as data from small fleets (instrumented vehicles) or with unclear methodology. Duplications were also removed. Finally, 74 sources were considered for the following review. Each of the topics listed in the Methodology is addressed below.

9.3.1 Sampling Rate

In the following text, a distinction will be made between two uses of FCD: either for navigation, or from a research/driver-focused perspective. Typically, FCD collected for navigation and traffic analyses (providing real-time traffic information, travel time predictions, etc.) is based on GNSS signals from vehicle fleets (taxis, commercial vehicles, and private vehicles using mobile phones or satellite navigation). For these purposes, a GNSS signal frequency of once every second is common. In many cases, data sample is aggregated and transmitted to the collection agency at longer intervals, as shown in Table 9.1.

Sampling rates have a direct impact on the available level of detail of obtained data. For example, 1 s (1 Hz) corresponds to approx. 14 and 25 m driven, at typical urban/rural speed limits of 50 and 90 km/h, respectively. This is why frequencies below 1 Hz (i.e., one or more records per second), are necessary for detailed studies. Higher frequency data enables detailed examination of what occurred just before, during and after a crash or safety critical event. In addition to speed data, acceleration data is of interest for certain road safety studies. Acceleration data is usually collected at higher frequencies than speed data, and used to derive various SMoS such as 'jerks', longitudinal harsh deceleration, harsh acceleration, harsh cornering, swerving, etc. The question is what minimum frequencies of speed and accelerometer data collection should be set for road safety investigations?

Table 9.1 FCD studies and their characteristics (sorted by sampling frequency)

References	Location, fleet	Sampling frequency
Berntsen et al. [8]	Telemotix (544 vehicles, Norway)	5 s
Wang et al. [9–11]	YOOTU (15,000 taxis, Shanghai)	10–15 s
Jurewicz et al. [12]	HERE and TomTom (Australia)	10–30 s
Bekhor et al. [13]	Decell (> 100,000 vehicles, Israel)	30 s
Hrubeš and Blümelová [14]	RODOS (> 100,000 vehicles, Czechia)	1 min
Pascale et al. [15]	WAY (13,000 trucks, Italy)	20 s–3 min
Aarts et al. [16]	TomTom (the Netherlands)	5 min

The current FCD market offers various sensors, capable of providing real-time GNSS and accelerometer data. However, the choice of frequency may influence the sample size; representativeness of the data; price of purchase; and data storage, transfer and processing. Ideally, data collection requirements should be planned according to the observed phenomenon. For example, in a NDS of motorcycle riders [17], GNSS sampling frequency was set at 100 Hz based on typical riders' reaction times of 0.3–0.4 s and the requirement of at least 15 signals for adequate instrumental description of rider reactions. However, most NDSs are not as strict: a review of such studies [18] listed typical GNSS and accelerometer sampling frequencies between 10 and 30 Hz. Another summary [19] recommended 50 Hz as sufficient for sampling acceleration data. In general, 10 Hz seems to be typical sampling rate for speed data for large NDS/FOT (e.g., 100-Car NDS, SHRP2 NDS, euroFOT, SeMiFOT).

On the other hand, for routinely collected data (not specifically planned for research), lower sampling frequencies are often used. For example, Bärgman [20] distinguished research data (often collected at 10 Hz or higher) and commercially collected data (usually with lower sample frequency, such as 4 Hz)—these may involve fleet monitoring or car insurers.

In general, usefulness of data depends on the purpose of individual safety studies and requirements to address specific research questions. These may comprise monitoring speed trends, informing strategy and doing evaluations at road segment level, as well as more detailed studies, using acceleration data. For selected examples of research studies, based on both data sources, see Table 9.2.

Apart from the mentioned GNSS and accelerometers, there are more sophisticated instrumented vehicles, which involve for example videos, VBOX sensors, Mobileye and/or LIDAR (see reviews [27, 28]). While they present excellent data acquisition systems for safety research, they are not likely to be feasible for large fleets due to their high cost. As large FCD fleets are necessary to provide large speed data sources for research consideration, these studies fall outside the scope of this review.

Table 9.2 Research-oriented FCD studies and their characteristics (sorted by sampling frequency)

References	Data provider (fleet, location)	Sampling frequency
Reinau et al. [21]	Denmark (ITS Platform)	Speed 1 Hz, acceleration 10 Hz
Toledo et al. [22]	Israel (GreenBox)	Acceleration 40 Hz
Ambros et al. [23]	Czechia (Princip)	Speed 4 Hz, acceleration 32 Hz
Bagdadi and Várhelyi [24]	Sweden (Lund ISA trial)	Jerk 5 Hz
Punzo et al. [25]	US (NGSIM program)	Jerk 10 Hz
Joubert et al. [26]	South Africa (Digicore)	Acceleration 50 Hz

9.3.2 *Study Size*

Conventional sampling theory calculates minimum sample size based on allowable error and sample standard deviation of measured speeds. Traditional recommendations were measuring at least 30, and ideally 100–200 vehicles (e.g., [29–32]). Also, a Transport Research Board synthesis of operating speed studies [32] reports a typical requirement of "at least 100 [vehicles] per site."

However, Smith et al. [33] suggested this approach is not fully transferable to FCD studies, where the conditions of traditional sampling theory (data within each measurement interval are stationary, and variance does not change) do not hold. On the contrary, FCD offer a non-constant vehicle fleet penetration rate. This raises the question of how large a FCD fleet sample is required to be representative of total traffic flow. Reviews summarized found in the highway environment penetration rates up to 3% are considered sufficient [34], and in urban areas rates up to 5% are recommended [2].

On roads with lower volumes, lack of data may be expected. Srinivasan and Jovanis [35] argued that "probes cannot be used as a stand-alone source of travel time information, especially during off-peak periods and on lightly travelled corridors and low-speed roads, such as local and collector streets and minor arterials". Nevertheless, recent expansion of FCD is changing this situation. Jurewicz et al. [12] studied FCD on lower-volume roads and provided some guidance on necessary data collection periods (in numbers of months, based on the level of traffic volume).

A potential limitation may be caused by the fact that fleets, from which FCD is sourced, may involve significant proportion of commercial vehicles and thus does not fully represent total traffic flow or driving behaviour. Commercial drivers may be also more likely to travel on familiar routes, major roads, etc. Analysts should consider this and possibly verify the representativeness of sample (fleet data) against population (total traffic flow).

To produce national/state safety performance indicators, the speed data should ideally come from the entire road network or the widest possible spectrum of locations to make it representative of the entire road network. A detailed manual of this process was developed by the EU SafetyNet project [36]. As a minimum, the sites should be sampled from sub-groups based on different road types, speed limits and/or number of lanes (approx. 30 sites per group). This could also serve as a minimum requirement for FCD-based speeds, if necessary.

9.3.3 *Free-Flow Speed Determination*

In traffic engineering, free-flow speed is often used as a standard measure, comarable across different sites and representing speed of vehicles under low volume conditions, unhindered by traffic control devices. It may be estimated using through spot-speeds, collected either automatically or manually. In this approach, free-flowing vehicles are selected individually by an observer. Another, more objective approach is based on gaps between vehicles—various headway or gap thresholds are used to distinguish between vehicles following others or travelling freely. However, many different values have been used in international guidance, ranging from 3 to 12 s. Other thresholds have been even more pragmatic: for example, in terms of hourly number of vehicles (ranging between studies from 200 to 1000 veh/hr) (for a review see [37]).

Nevertheless, none of these approaches is feasible for FCD, which is collected from individual vehicles only, without being able to check whether they are influenced by other vehicles or not. One approach is restricting data collection to off-peak hours [13, 38, 39], or night-time [40]. However, this practice is likely to severely reduce the sample, especially in case of commercial vehicles, which usually travel during daytime. One could also ask whether night-time speeds are representative of typical driver behaviour, as these speeds may be influenced by darkness, fleet composition, presence/absence of street lighting, and potentially increased speeding [12]. The question could be even more general: if we aim to study unsafe driving behaviour, should we focus on speed data from the times and conditions when crashes happen most? If so, then it may not make sense to study free-flow conditions, where vehicles are not affected by presence of other road users.

In general, free-flow speed issues have not been studied much in FCD literature. For example, Diependaele et al. [41] attempted modelling the free-flow speeds with a probabilistic approach, and Ambros et al. [23] applied cluster analysis to separate free-flow speeds from all-vehicle speed data. For practical purposes, FCD-based speeds from off-peak periods may be assumed to be an estimate of free-flow speeds, with acknowledgement of the limitation that these may not be representative of all vehicles or driving behaviour. The specific selection of these periods may need to be judgement-based.

When using FCD to derive free flow speeds, it is important to make sure that speeds are:

- reliable when compared between various providers,
- valid when compared to "ground truth" (traditional speed measurement methods),
- relevant to road safety.

9.3.4 Reliability

Spasovic et al. [42] reported a US study assessing the reliability of FCD-based speeds provided by three commercial traffic data providers (INRIX, NAVTEQ, TraficCast Dynaflow), using data from four roadways in New Jersey and New York. All three technologies were mostly within the acceptance limits for the average absolute speed error ($\leq$16 km/h) and the speed error bias ($\leq$8 km/h). All of the studied technologies consistently overestimated the speed in the lowest speed bin (0–50 km/h), and underestimated the speed in the highest speed bin (>100 km/h).

Another US study [43] investigated whether there is a relationship between HERE, INRIX and Bluetooth speed data. Modelled free-flow speeds were found on an average 10% higher than the observed Bluetooth speeds; Bluetooth and HERE travel speeds were in general 8–16 km/h lower than INRIX speeds during the day.

9.3.5 *Validity*

Several comparative studies investigated the validity of FCD-based travel times and speeds against traditional data sources: see Table 9.3. Due to the wide range of study types, FCD sources, and differences in their robustness, only high-level findings are summarized in Table 9.3 to provide a general overview of validity.

Some of the mentioned studies found FCD-based speeds were higher than other measures of speed; other studies [12, 16] found an opposite tendency (speed from FCD lower than other measures of speed). This may be because FCD-based average speed relates to a whole road segment (i.e., including turning at intersections), while the traditional spot-speed relates to a single spot only (typically collected away from intersections [16]) or specific road lengths, such as point-to-point speed measures. Also, different FCD sources and frequency of detection were used in the studies leading to different outcomes. It is important to understand the specifications of each FCD source and to check or calibrate its speed outputs against a trusted ground truth source.

To sum up, many studies concluded that FCD-based speed is valid (not more than by 16 km/h different compared to ground truth or other providers' data). For research purposes, this difference may be too large; if the differences are systematic, consideration should be given to treating or calibrating the collected FCD to align the data with the ground truth on case-by-case basis.

9.3.6 *Use Cases*

Information distilled from FCD studies has the potential to enhance and improve the quality and coverage of speed and safety studies. Three following use cases were selected as example applications.

Use case 1: FCD-based speeds may be used network-wide as a safety performance indicator. This was an idea of a Dutch analysis reported by Aarts et al. [16]. After investigating performance of TomTom data, the source was found feasible for providing information for safety performance indicators (specifically speed levels). The study noted some limitations of FCD, for example they do not provide information about speed differences between surrounding vehicles; privacy issues also may limit analyses related to vehicle types or driver age/gender. On the other hand, in an Australian study, Jurewicz et al. [12] found FCD could be potentially translated to

Table 9.3 FCD-based comparative studies and their characteristics (sorted by study date)

References	Data description	High-level findings
Travel time studies		
Clergue and Buttignol [44]	4 routes in France (penetration 0.7–4.3%), TomTom versus ALPR	"Differences are insignificant"
de Boer and Krootjes [45]	9 routes in Eindhoven (the Netherlands), penetration >2%, TomTom historic travel times versus automated license plate recognition (ALPR) for point-to-point speed detection	"FCD is accurate"
Brockfeld et al. [46]	4 days of data from 500-taxi FCD fleet in Nuremberg (Germany) versus ALPR	"Travel times calculated by the system deliver valuable data"
Speed studies		
Jurewicz et al. [12]	Roads in Victoria, Australia, FCD versus loops	FCD speed lower, with standard error 9.7 km/h
Ambros et al. [23]	Czech rural roads, FCD versus radar	FCD speed on average 2 km/h *higher*
Diependaele et al. [41]	Belgian rural roads, FCD versus loops	FCD speed on average almost 10 km/h *higher* than free-flow loop-speed
INRIX [47]	"The World's Largest Independent Traffic Data Validation" (I-95 VPP)	FCD speed accurate within 16 km/h of actual traffic speeds on average
Hrubeš and Blümelová [14]	Prague ring road (Czech Republic), 2% penetration, FCD versus loops	FCD speed "reasonable estimate of speed", shown to be generally *lower*
Espada and Bennett [48]	EastLink in Melbourne, HERE travel speeds every 5 min versus e-tag gate crossings	HERE probe travel speed estimates on average 9 km/h *lower*
Lattimer and Glotzbach [49]	INRIX FCD versus ALPR data on 4 Florida freeways	INRIX speeds on average 10 km/h *higher*
Bar-Gera [50]	Cellular FCD versus dual magnetic loop detectors (Israeli freeway)	"A good match between the two measurement methods"
Smith et al. [33]	Cellular FCD (10-min intervals) versus point video	FCD on average 10–15 km/h *higher*
Yim [51]	Cellular phone-based speeds versus loop speeds over 1 month (four French freeways)	Cellular data speeds about 10% *lower* on intercity freeways, and 24–32% *higher* on an urban freeway

spot-speed equivalent using calibration models; the authors also provided an example of using FCD-based speeds for before-after evaluation of a speed limit change.

Use case 2: FCD may be used to derive speeding or harsh braking/accelerating to identify dangerous events and assess driving styles. There is evidence that some safety critical event algorithms related to speed and acceleration are predictive of crash involvement risk [52]. For instance, various measures of deceleration have been found to be associated with increased crash or conflict/near-crash frequency (e.g., [53–56]). Additionally, deceleration measures have been associated with behaviours that may contribute to increased crash risk. Rapid deceleration events (RDEs) have been successfully used as a SMoS in studies of older driver safety [57–59]. In this regard, FCD linked to specific drivers presents a valuable source for assessing driving performance and driving styles, as well as driving exposure. This data may then be employed for usage-based insurance systems [60]. However, large amounts of collected data need to be analyzed, and there is no guideline in determining necessary amount of data [61]. Ellison et al. [62] mentioned various approaches to reduce this burden, such as using pattern matching algorithms to identify patterns of interest and focus analysis on these portions of the data, including verification through video footage. Furthermore, FCD may be influenced by exogenous factors, such as congestion, construction, traffic light timings and other vehicles. The use of verification through review of video footage may reduce these influences, but also requires labour intensive manual processing. Fortunately, there are approaches to recognize conflicts within FCD without manually reviewing all video streams. A common approach is to analyze kinematic vehicle data to detect safety-critical events. However, critical values (thresholds) of these "event triggers" vary significantly in the literature [63], for example:

- longitudinal deceleration ranges from approx. 0.1 to 0.75 g [54, 58, 64–70].
- critical jerks vary between 0.06 and 2 g/s (e.g., [71, 72]),

Safety critical events based on combined criteria have also been used by some researchers (e.g., [71, 73]). An alternative approach is analysing all the collected data (so called risk space [26]). Note that smartphones are often used for collecting data for driver assessment. However, studies indicate smartphones may not be as suitable for recording in-vehicle driving data as in-vehicle installed data collection systems. Paefgen et al. [69] found smartphone FCD overestimated critical driving events, and Händel et al. [74] reported the data lacked sufficient reliability to accurately determine safety critical events. This may need to be reviewed with constant technology improvements within smartphones [75].

Use case 3: FCD may be used to obtain speed and other indicators to identify and assess safety at specific locations.

Based on the meaningful risk thresholds and frequency of their occurrence at specific locations, high-risk sites may be identified. Reviewed studies examining this are summarised in Table 9.4.

Collecting network-wide FCD also enable studying relationships between speed and driving/environmental characteristics. For example, in Denmark, using FCD linked with relevant data from an ITS Platform enabled quantifying the influence of

Table 9.4 Validation approaches in selected FCD-based studies (sorted by study date)

References	Location; indicator type	Validation
Kamla et al. [67]	UK roundabouts; acceleration	Significant impact of frequency of truck harsh braking incidents (>4.4 m/s^2) in an accident prediction model
Stipancic et al. [76]	Quebec City; acceleration	Correlation between critical events (±2 m/s^2) and crash frequencies
Pande et al. [72]	California freeways; jerks	Relating 10 jerk thresholds (varying from 0.50 to 2.75 ft/s^3, with an increment of 0.25) to crash frequency
Ambros et al. [23]	Czech rural roads; speed	Speed consistency (i.e., differences between speeds in tangents and following curves) related to a long-term crash frequency
Reinau et al. [21]	Aalborg city (Denmark); speed and jerks	Visual comparison of crash location map versus risk location map
Mousavi et al. [77]	Louisiana highways; jerks	21 jerk value thresholds evaluated in the sensitivity analysis; segment jerk-rates compared to crash rates

road and shoulder width, curve radii, the extent of road markings and the section lengths on speed [78]. A model, based on Czech FCD on rural roads, confirmed that increasing road width and enabling overtaking and climbing are associated with an increase of speed [23]. An Israeli FCD study [79] found changing shoulder width or recovery-zone width (clear zone) has the potential to affect driving speeds.

Unfortunately, in several studies the explanatory power (R2) of the mentioned FCD-based speed models was relatively low (approx. 30–40% [23, 78, 80]). This finding may be explained by the characteristics of FCD: conventional spot-speed data is based on samples collected in more or less controlled conditions (daytime, season, weather, etc.) and may thus yield homogeneous results with high R2 values. On the other hand, FCD studies use an "anonymous" sample collected in various conditions, which may lead to heterogeneous results with lower R2 values. The low explanatory power may lead to insufficient reliability in cases when models are applied in different time and space from the original conditions. Therefore, the models could benefit from improvement: for example, potentially by adding additional explanatory variables, and/or considering vehicle and driver characteristics using random effect models [81].

9.4 Summary, Discussion and Conclusions

The goal of this review was to identify challenges and opportunities regarding using FCD to develop potential speed-related SMoS. However, the reviewed studies varied in purpose and quality, which limited comparability beyond high-level findings.

Robust studies were thus given more prominence. In addition, as the reviewed field is quickly evolving, new studies are being published and may change the validity of the reported findings. Regarding conclusions, firstly it is important to consider the benefits and limitations of FCD:

- compared to traditional spot-speed measurements, FCD is able to provide unlimited spatial coverage, as well as historical data. However, FCD may not be sufficient in the case of low traffic volumes. Representativeness may also be influenced by character of the fleet (often mostly commercial vehicles),
- anonymity of FCD may limit distinguishing different vehicle types or driver characteristics. It also complicates the determination of free-flow speed,
- continuity and quality of FCD measurements is beyond the direct influence of end users.

Secondly, it is important to remember FCD originally served navigation and traffic monitoring purposes. To ensure FCD may be confidently used in road safety research, the differences from the original purposes need to be considered in the context of SMoS:

- sampling rate needs to be known and planned, based on requirements and type of data collected,
- study size also needs to be planned, especially in conditions of low traffic volume,
- there is no universal approach to estimating free-flow speeds,
- reliability and validity: FCD-based speed reliability and relation to the ground truth is uncertain and strongly dependent on the compared sources. It should be tested for each new fleet/source,
- there are no guidelines for determining SMoS algorithms or thresholds (e.g., rapid braking), nor uniform approaches to validating FCD-based speeds against safety.

Some issues may be inherent to the method: for example, FCD is usually collected in urban areas, with a not-fully-representative vehicle fleet and driver sample. Both pros and cons of FCD need to be carefully weighed, based on the requirements of specific research tasks.

Nevertheless, FCD quality and coverage is continuously increasing. FCD found its way into commercial services, such as PTV Visum (with TomTom FCD) or VIA Traffic Solutions Software and ARRB Aperture tool (with HERE Traffic Analytics). Continued data collection and investigations focusing on the mentioned issues will assist with developing effective FCD speed-derived SMoS.

Acknowledgements The chapter was produced with the financial support of Czech Ministry of Education, Youth and Sports under the National Sustainability Program I project of Transport R&D Centre (LO1610), using the research infrastructure from the Operational Program Research and Development for Innovation (CZ.1.05/2.1.00/03.0064).

References

1. Chang A, Saunier N, Laureshyn A (2017) Proactive methods for road safety analysis. SAE International, Warrendale
2. Bessler S, Paulin T (2013) Literature study on the state of the art of probe data systems in Europe. FTW Telecommunications Research Center, Vienna
3. Nilsson G (2004) Traffic safety dimensions and the Power Model to describe the effect of speed on safety. Lund University, Lund
4. Elvik R (2009) The Power Model of the relationship between speed and road safety: Update and new analyses. Institute of Transport Economics, Oslo
5. Elvik R (2013) A reparameterisation of the Power Model of the relationship between the speed of traffic and the number of accidents and accident victims. Accid Anal Prev 50:854–860
6. Leduc G (2008) Road traffic data: collection methods and applications. European Communities, Luxembourg City
7. Rose G (2006) Mobile phones as traffic probes: Practices, prospects and issues. Transp Rev 26:275–291
8. Berntsen M, Molnár P, Zdĕnek M (2016) Driving behavior and risk of accident: evidence from Norway. Norwegian University of Science and Technology, Trondheim
9. Wang X, Fan T, Chen M, Deng B, Wu B, Tremont P (2015) Safety modeling of urban arterials in Shanghai, China. Accid Anal Prevent 83:57–66
10. Wang X, Fan T, Li W, Yu R, Bullock D, Wu B, Tremont P (2016) Speed variation during peak and off-peak hours on urban arterials in Shanghai. Transp Res Part C 67:84–94
11. Wang X, Zhou Q, Quddus M, Fan T, Fang S (2018) Speed, speed variation and crash relationships for urban arterials. Accid Anal Prev 113:236–243
12. Jurewicz C, Espada I, Makwasha T, Han C, Alawi H, Ambros J (2017) Validation and applicability of floating car speed data for road safety. In: Australasian road safety conference, Perth
13. Bekhor S, Lotan T, Gitelman V, Morik S (2013) Free-flow travel speed analysis and monitoring at the national level using global positioning system measurements. J Transp Eng ASCE 139:1235–1243
14. Hrubeš P, Blümelová J (2015) Comparative analysis for floating car and loop detectors data. In: 22nd ITS World Congress, Bordeaux
15. Pascale A, Deflorio F, Nicoli M, Dalla Chiara B, Pedroli M (2015) Motorway speed pattern identification from floating vehicle data for freight applications. Transp Res Part C 51:104–119
16. Aarts LT, Bijleveld FD, Stipdonk HL (2015) Usefulness of 'floating car speed data' for proactive road safety analyses: analysis of TomTom speed data and comparison with loop detector speed data of the provincial road network in the Netherlands. SWOV Institute for Road Safety Research, Leidschendam
17. Laporte S (ed) (2010) Design of a naturalistic riding study-implementation plan. 2BESAFE project deliverable 5
18. Backer-Grøndahl A, Phillips R, Sagberg F, Touliou K, Gatscha M (2009) Topics and applications of previous and current naturalistic driving studies. PROLOGUE project deliverable 1.1
19. Welsh R, Reed S, Talbot R, Morris A (2010) Data collection, analysis methods and equipment for naturalistic studies and requirements for the different application areas. PROLOGUE project deliverable 2.1
20. Bärgman J (2015) On the analysis of naturalistic driving data: development and evaluation of methods for analysis of naturalistic driving data from a variety of data sources. Chalmers University of Technology, Gothenburg
21. Reinau KH, Andersen CS, Agerholm N (2016) A new method for identifying hazardous road locations using GPS and accelerometer. In: 23rd ITS World Congress, Melbourne
22. Toledo T, Musicant O, Lotan T (2008) In-vehicle data recorders for monitoring and feedback on drivers' behavior. Transp Res Part C 16:320–331

23. Ambros J, Valentová V, Gogolín O, Andrášik R, Kubeček J, Bíl M (2017) Improving the self-explaining performance of Czech national roads. Transp Res Rec 2635:62–70
24. Bagdadi O, Várhelyi A (2011) Jerky driving—an indicator of accident proneness? Accid Anal Prev 43:1359–1363
25. Punzo V, Borzacchiello MT, Ciuffo B (2011) On the assessment of vehicle trajectory data accuracy and application to the Next Generation SIMulation (NGSIM) program data. Transp Res Part C 19:1243–1262
26. Joubert JW, de Beer D, de Koker N (2016) Combining accelerometer data and contextual variables to evaluate the risk of driver behaviour. Transp Res Part F 41:80–96
27. Carsten O, Kircher K, Jamson S (2013) Vehicle-based studies of driving in the real world: The hard truth? Accid Anal Prev 58:162–174
28. Valero-Mora PM, Tontsch A, Welsh R, Morris A, Reed S, Touliou K, Margaritis D (2013) Is naturalistic driving research possible with highly instrumented cars? Lessons learnt in three research centres. Accid Anal Prev 58:187–194
29. Kraft WH, Homburger WS, Pline JL (eds) (2019) Traffic engineering handbook, 6th edn. Institute of Transportation Engineers, Washington
30. Narasimha Murthy AS, Mohle HR (2001) Transportation engineering basics, 2nd edn. American Society of Civil Engineers, Reston
31. PIARC (2003) Road Safety Manual: recommendations from the World Road Association (PIARC). Route2market, Harrogate
32. Transportation Research Board (2011) Modeling operating speed: synthesis report. Transportation Research Board, Washington
33. Smith BL, Zhang H, Fontaine M, Green M (2003) Cell-phone probes as an ATMS tool. Smart Travel Laboratory, Charlottesville
34. Vandenberghe W, Vanhauwaert E, Verbrugge S, Moerman I, Demeester P (2012) Feasibility of expanding traffic monitoring systems. IET Intel Transport Syst 6:347–354
35. Srinivasan KK, Jovanis PP (1996) Determination of number of probe vehicles required for reliable travel time measurement in urban network. Transp Res Rec 1537:15–22
36. Hakkert AS, Gitelman V (eds) (2007) Road safety performance indicators: manual. SafetyNet project deliverable 3.8
37. Ambros J, Kyselý M (2016) Free-flow vs car-following speeds: Does the difference matter? Adv Transp Studies 40:17–26
38. Pline JL (ed) (1992) Traffic engineering handbook, 4th edn. Institute of Transportation Engineers, Washington
39. Wang J, Dixon K, Li H, Hunter M (2006) Operating-speed model for low-speed urban tangent streets based on in-vehicle global positioning system data. Transp Res Rec 1961:24–33
40. TomTom traffic index: measuring congestion worldwide. https://www.tomtom.com/en_gb/trafficindex/about
41. Diependaele K, Riguelle F, Temmerman P (2016) Speed behavior indicators based on floating car data: results of a pilot study in Belgium. Transportation Research Procedia 14:2074–2082
42. Spasovic LN, Dimitrijevic B, Kim K (2013) Probe vehicle data comparative validation study—New Jersey and New York—final report. New Jersey Institute of Technology, Newark
43. Rapolu S, Kumar A (2015) Comparing arterial speeds from "big-data" sources in Southeast Florida (Bluetooth, HERE and INRIX). In: 15th TRB national transportation planning applications conference, Atlantic City
44. Clergue L, Buttignol V (2014) Using GPS data in favour of traffic knowledge. In: Transport Research Arena, Paris
45. De Boer G, Krootjes P (2012) The quality of floating car data benchmarked: an alternative to roadside equipment? In: 19th ITS World Congress, Vienna
46. Brockfeld E, Lorkowski S, Mieth P, Wagner P (2007) Benefits and limits of recent floating car data technology—an evaluation study. In: 11th WCTR Conference, Berkeley
47. INRIX I-95 VPP data summary validation. http://inrix.com/case-studies/inrix-i-95-vpp-data-summary-validation-case-study/

48. Espada I, Bennett P (2015) Probe data and its application in traffic studies. In: IPWEA/IFME conference, Rotorua
49. Lattimer CR, Glotzbach G (2012) Evaluation of third-party travel time data in Tallahassee. FL. In ITS America, National Harbor
50. Bar-Gera H (2007) Evaluation of a cellular phone-based system for measurements of traffic speeds and travel times: a case study from Israel. Transp Res Part C 15:380–391
51. Yim Y (2003) The state of cellular probes. University of California, Berkeley
52. Sagberg F, Selpi S, Piccinini GFB, Engström J (2015) A review of research on driving styles and road safety. Hum Factors 57:1248–1275
53. Dingus TA, McGehee DV, Manakkal N, Jahns SK, Carney C, Hankey JM (1997) Human factors field evaluation of automotive headway maintenance/collision warning devices. Hum Factors 39:216–229
54. Feng F, Bao S, Sayer JR, Flannagan C, Manser M, Wunderlich R (2017) Can vehicle longitudinal jerk be used to identify aggressive drivers? An examination using naturalistic driving data. Accid Anal Prev 104:125–136
55. Kiefer RJ, Flannagan CA, Jerome CJ (2006) Time-to-collision judgments under realistic driving conditions. Hum Factors 48:334–345
56. Markkula G, Engström J, Lodin J, Bärgman J, Victor T (2016) A farewell to brake reaction times? Kinematics-dependent brake response in naturalistic rear-end emergencies. Accid Anal Prevent 95:209–226
57. Chevalier A, Chevalier AJ, Clarke E, Coxon K, Brown J, Rogers K, Boufous S, Ivers R, Keay L (2016) Naturalistic rapid deceleration data: drivers aged 75 years and older. Data Brief 9:909–916
58. Chevalier A, Coxon K, Chevalier AJ, Clarke E, Rogers K, Brown J, Boufous S, Ivers R, Keay L (2017) Predictors of older drivers' involvement in rapid deceleration events. Accid Anal Prev 98:312–319
59. Keay L, Munoz B, Duncan DD, Hahn D, Baldwin K, Turano KA, Munro CA, Bandeen-Roche K, West SK (2013) Older drivers and rapid deceleration events: Salisbury eye evaluation driving study. Accid Anal Prev 58:279–285
60. Tselentis DI, Yannis G, Vlahogianni EI (2017) Innovative motor insurance schemes: a review of current practices and emerging challenges. Accid Anal Prev 98:139–148
61. Stavrakaki A-M, Tselentis DI, Barmpounakis EN, Vlahogianni EI, Yannis G (2019) How much driving data do we need to assess driver behavior? In: 98th transportation research board annual meeting, Washington
62. Ellison AB, Greaves SP, Bliemer MCJ (2015) Driver behaviour profiles for road safety analysis. Accid Anal Prev 76:118–132
63. Aichinger C, Nitsche P, Stütz R, Harnisch M (2016) Using low-cost smartphone sensor data for locating crash risk spots in a road network. Transp Res Procedia 14:2015–2024
64. Baldwin KC, Duncan DD, West SK (2004) The driver monitor system: a means of assessing driver performance. Johns Hopkins APL Technical Digest 25:269–277
65. Hickman JS, Hanowski RJ (2011) Use of a video monitoring approach to reduce at-risk driving behaviors in commercial vehicle operations. Transp Res Part F 14:189–198
66. Johnson DA, Trivedi MM (2011) Driving style recognition using a smartphone as a sensor platform. In: 14th international IEEE conference on intelligent transportation systems, Washington (2011)
67. Kamla J, Parry T, Dawson A (2019) Analysing truck harsh braking incidents to study roundabout accident risk. Accid Anal Prev 122:365–377
68. Musicant O, Botzer A, Laufer I, Collet C (2018) Relationship between kinematic and physiological indices during braking events of different intensities. Hum Factors 60:415–427
69. Paefgen J, Kehr F, Zhai Y, Michahelles F (2012) Driving behavior analysis with smart-phones: insights from a controlled field study. In: 11th international conference on mobile and ubiquitous multimedia, Ulm
70. Ryder B, Gahr B, Egolf P, Dahlinger A, Wortmann F (2017) Preventing traffic accidents with in-vehicle decision support systems—the impact of accident hotspot warnings on driver behaviour. Decis Support Syst 99:64–74

71. Naude C, Serre T, Dubois-Lounis M, Fournier J-Y, Lechner D, Guilbot M, Le-doux V (2019) Acquisition and analysis of road incidents based on vehicle dynamics. Accid Anal Prev 130:117–124
72. Pande A, Chand S, Saxena N, Dixit V, Loy J, Wolshon B, Kentda JD (2017) A preliminary investigation of the relationships between historical crash and naturalistic driving. Accid Anal Prev 101:107–116
73. Dingus TA, Klauer SG, Neale VL, Petersen A, Lee SE, Sudweeks JD, Perez MA, Hankey J, Ramsey DJ, Gupta S, Bucher C, Doerzaph ZR, Jermeland J, Knipling RR (2006) The 100-car naturalistic driving study, phase II—results of the 100-car field experiment. National Highway Traffic Safety Administration, Washington
74. Händel P, Skog I, Wahlström J, Bonawiede F, Welch R, Ohlsson J, Ohlsson M (2014) Insurance telematics: opportunities and challenges with the smartphone solution. IEEE intelligent transportation systems magazine, pp 57–70
75. Ziakopoulos A, Tselentis D, Kontaxi A, Yannis G (2020) A critical overview of driver recording tools. J Safety Res 72:203–212
76. Stipancic J, Miranda-Moreno L, Saunier N (2018) Vehicle manoeuvers as surrogate safety measures: Extracting data from the GPS-enabled smartphones of regular drivers. Accid Anal Prev 115:160–169
77. Mousavi S-M, Parr SA, Pande A, Wolshon B (2015) Identifying high-risk roadways through jerk-cluster analysis. In: 2015 road safety and simulation international conference, Orlando
78. Andersen CS, Reinau KH, Agerholm N (2016) The relationship between road characteristics and speed collected from floating car data. J Traffic Transp Eng 4:291–298
79. Gitelman V, Pesahov F, Carmel R, Bekhor S (2016) The identification of infrastructure characteristics influencing travel speeds on single-carriageway roads to promote self-explaining roads. Transp Res Procedia 14:4160–4169
80. Gaca S, Kieć M (2016) Speed management for local and regional rural roads. Transp Res Procedia 14:4170–4179
81. Bassani M, Cirillo C, Molinari S, Tremblay J (2016) Random effect models to predict operating speed distribution on rural two-lane highways. J Transp Eng ASCE 142:04016019

Chapter 10
Unsafe Driving Behaviours at Single-Lane Roundabouts: Empirical Evidence from CHAID Method

Natalia Distefano, Giulia Pulvirenti, Salvatore Leonardi, and Tomaž Tollazzi

10.1 Introduction

Even though roundabouts comprise just a small amount of the roadway network, they generally are more complex and difficult to navigate than most other road segments. To drive in a roundabout properly and safely, the driver has to take into account the layout of the site, its geometry, and interactions with other users. Roundabout can be geometrically complex and it requires that drivers scan several different areas and keep track of several different elements to get the right information needed to perform the correct manoeuvre. So, while driving roundabouts, drivers have to perform many subtasks in an environment with many rules and complex interactions. Each task encountered by the driver at roundabout involves a sequence of: perception or recognition; decision making; execution or performance; and real time system response by the vehicle, roadway and surrounding environment. From a cognitive analysis perspective, the interaction between the car drivers' capabilities and the demands of the driving task determines the outcome in terms of a more or less safe driving behaviour [1].

N. Distefano (✉) · G. Pulvirenti · S. Leonardi
Department of Civil Engineering and Architectural, University of Catania, Catania, Italy
e-mail: natalia.distefano@unict.it

G. Pulvirenti
e-mail: giulia.pulvirenti@unict.it

S. Leonardi
e-mail: salvatore.leonardi@unict.it

T. Tollazzi
Faculty of Civil Engineering, Transportation Engineering and Architecture, University of Maribor, Maribor, Slovenia
e-mail: tomaz.tollazzi@um.si

E. Macioszek and G. Sierpiński (eds.), *Research Methods in Modern Urban Transportation Systems and Networks*, Lecture Notes in Networks and Systems 207, https://doi.org/10.1007/978-3-030-71708-7_10

Previous studies show that the drivers' driving behaviour at roundabout changes depending on the layout of the roundabout [2, 3]. Other studies [4, 5] modelled speed profiles at roundabouts and concluded that speed profiles differed significantly across drivers and roundabouts. A recent study shows that the stress level induced by roundabouts on drivers is more than double that induced by standard intersections [6].

The understanding of the correlations among driver behaviour and driver characteristics and road environment is very important in order to help drivers to make safe manoeuvres in increasingly complex intersections.

Many factors contribute to road safety. Some involve planning, design, construction, operation, and policing of the roadways. The most relevant factor is human factor. This includes unawareness of traffic rules and roadway condition; lack of driving skills; poor judgment; failure to interact and adjust to prevailing roadway conditions; and most importantly, aggressive driving [7].

The "incorrect" design of infrastructure (for example inappropriately designed road, wrong layout of intersections, poor positioning of pedestrian crossings, poor side distance, etc.) causes wrong driver's behaviour and other subsequent problems, i.e. a conflict situation or a traffic accident [7–9].

Previous crash investigations suggested that human error was a contributing factor to the accidents. Where human factor is cited as a contributing factor it is important to seek out where the failures occur in order to find appropriate treatments to reduce the likelihood of the same event occurring again.

This chapter aims to investigate the nature and sources of unsafe driving behaviour at single-lane roundabouts using the results of driving experiment. In fact, it is believed that the possible user's unsafe behaviour could be due to critical geometric configuration of the roundabouts.

In this study, we attempt to define unsafe driving behaviour as the complex of deliberate and systematic practices that increase the risk of a conflict or crash and also investigate the types of road users and geometric characteristics of roundabouts that are directly associated with these behaviours.

These behaviours (hereinafter referred to as unsafe behaviour) represent driving activities that are often linked to crashes on the road. According to [10], these types of behaviours have the potential to degrade driving performance resulting in serious consequences for road safety and in addition greatly increase the risk of crashes.

This study has provided the favourable opportunity of observing what happens before most crashes occur. This is important because it is a proactive approach to traffic safety analysis without necessarily waiting for crashes to happen.

10.2 Literature Review

Several researches have demonstrated the substantial contribution of roundabouts in the enhancement of safety and efficiency while comparing it with various other intersection types [11–13]. The fundamental design of roundabouts is inclusive of the geometric layout, operational as well as safety evaluation. Small changes in roundabout geometry cause significant changes in its performance related to safety and operations [14].

Despite the proven safety benefits, some crashes still occur at roundabouts. Previous studies have examined roundabouts crashes and identified major crash types and contributing factors [15, 16]. Contributing factors to roundabouts crashes included entering drivers failing to yield the right-of-way to circulating vehicles, unsafe speeds, and incorrect lane choices at multilane roundabouts [17].

Alshannaq and Imam [7] found that all of the causes of roundabout crashes were related to the drivers' behaviour one way or another. Factors such as age, judgment, driving skills, attention, fatigue, experience, etc. were all found to be contributing factors to the occurrence of accidents. Similar findings have been reported in Ramisetty-Mikler et al. [18], which focused on the risky driving behaviour of Saudi Arabian adolescents. This study also highlighted factors such as young age, deficiency in training, and poor driving skills contributing to vehicle accidents. Bener et al. [19] by means of a study on driver behaviour in Qatar and Turkey demonstrated that the drivers' socio-economic conditions, driving style and skills, cultural factors, education, as well as ethnicity, contribute to traffic rule violations. Al-Rukaibi et al. [20] quantified the driver's behaviour at a roundabout. It was found that a large percent of drivers violates the traffic control devices. AL-Saleh and Bendak [21] performed another study in Al-Riyadh analysing the driver behaviour at roundabouts. They found that 90% of the chosen sample breached at least one traffic regulation. Distefano et al. [22] show that drivers adapt their driving behaviour according to their preferences on the geometric characteristics of roundabouts.

In recent times researchers have made a strong link between road safety and road user misbehaviour [8, 23–25]. For a long time, human error was most often considered as the main and more or less fatal cause of road safety problems since humans are, by nature, subject to errors.

The term human error has been used rather loosely to encompass nearly all the ways in which people can contribute to accidents through the performance of unsafe acts. Recent analyses, however, have shown that unsafe acts (i.e., potentially dangerous actions carried out in hazardous conditions) can be subdivided into two distinct classes of behaviour: errors and violations [26]. Errors have been defined in a variety of ways. In this context, an error is defined as the failure of planned actions to achieve their desired outcome without the intervention of some chance or unforeseeable agency. Violations, on the other hand, may be defined as the deliberate infringement of some regulated or socially accepted code of behaviour. Errors and violations differ both in their psychological mechanisms and in the kinds of remedial actions necessary to combat them. Errors arise as the result of information-processing problems; violations have a large motivational component [26].

Despite unsafe driving behaviours occur at all points in the road network, this may be particularly prominent at intersections, as these represent a complex part of the road system, where drivers are required to make decisions often within small timeframes. There is therefore a pressing need to examine the nature of drivers' unsafe driving behaviour at intersections, including the factors that contribute to, or mitigate these behaviours from occurring [24].

Drivers' unsafe driving behaviours are influenced by a range of personal, environmental or infrastructure factors. These factors include: inadequate knowledge, skills, training, wilful inappropriate behaviour, infrastructure and environment problems [27].

Mandiartha et al. [28], hypothesised that the geometric design and layout influence road users' behaviour. Average speeds of users were compared to test the hypothesis by using two different roundabout layouts. Although the roundabouts shared similar traffic volume, the behaviour of the users differed. The results of a study of Papantoniou et al. [29] reveal that the impact of driver characteristics and area type are the only statistically significant factors affecting the probability of driving errors. Other research found that drivers who report more experience with roundabouts also are more likely to know how to correctly navigate them, therefore driver behaviour at roundabouts has been observed to change with driver age [30–32].

Previous studies show that there is a significant need for identifying drivers who engage in unsafe driving practices, placing themselves and other road users at greater risk of involvement in a crash. This in order to design the roundabout layout such that deviant behaviour is not viable in physical terms.

10.3 Study Methodology

10.3.1 Participants

Sixty-six drivers (41 males and 25 females) aged 18–65 years took part in the study. All participants held a full valid license. To vary age, they were grouped into younger drivers (18–25 years), middle-aged drivers (26–50 years) and older drivers (51–65 years). Participants' characteristics are presented in Table 10.1.

Table 10.1 Features of participants

Category	Number	Percentage
Age		
18–25	22	33.33
26–50	26	39.39
51–65	18	27.28
Gender		
Male	41	62.12
Female	25	37.88

10.3.2 Test Route

A 30 km urban route around the suburbs surrounding the city of Catania was used for the on-road test. The urban route was composed of two routes: test route 1, which is 16 km long and contains 9 roundabouts (roundabouts 1–9), and test route 2, which is 14 km long and contains 8 roundabouts (roundabouts 10–17). Participants made the two routes at different times to avoid that the fatigue related to the excessive length of the route could affect their driving behaviour. The roundabouts were chosen in order to have different geometric characteristics. 13 roundabouts were single-lane, while 4 roundabouts were double-lanes. Four types of manoeuvres were considered: 1st exit (i.e. the driver took the 1st exit of the roundabout); 2nd exit (i.e. the driver took the 2nd exit of the roundabout); 3rd exit (i.e. the driver took the 3rd exit of the roundabout); U-turn (i.e. the driver exit from the same leg he/she was entered). For each roundabout the drivers made at least one manoeuvre. For this study only the drivers' unsafe behaviours made on single-lane roundabouts were considered. This study regards therefore 13 roundabouts (i.e. roundabouts 1 to 10 and roundabouts 15 to 17). Figure 10.1 shows the two test routes and the 13 one-lane roundabouts considered for this study. Data for roundabouts 11, 12, 13 and 14 (double-lane roundabouts) will be analysed in future research.

10.3.3 Experiment Design

Two in-vehicle observers used a behaviour pro-forma to manually record the unsafe behaviours made during the drive. On-route, the observer located in the front passenger seat provided directions. Both observers recorded unsafe behaviours made by the driver throughout the drive, including the type, where on the route it occurred and the context in which it occurred. Upon completion of the drive, the two observers checked agreement on the unsafe behaviours recorded.

The driving experiments were conducted during off-peak hours with low traffic volume. All testing was conducted during 9:30 am and 10:30 am and 14:30 and 15:30 pm on weekdays during September and October 2019. Each participant made the test on both test route 1 and 2, in different days.

10.3.4 Drivers' Unsafe Behaviours Classification

The unsafe behaviours observed during the on-road study were classified post hoc into specific types that directly reflected the nature of the behaviour made. The specific unsafe behaviours types were subdivided into three different categories: unsafe behaviours made during the approach to the roundabout (entry unsafe behaviours), unsafe behaviours made during the circulation on the roundabout (circulation unsafe

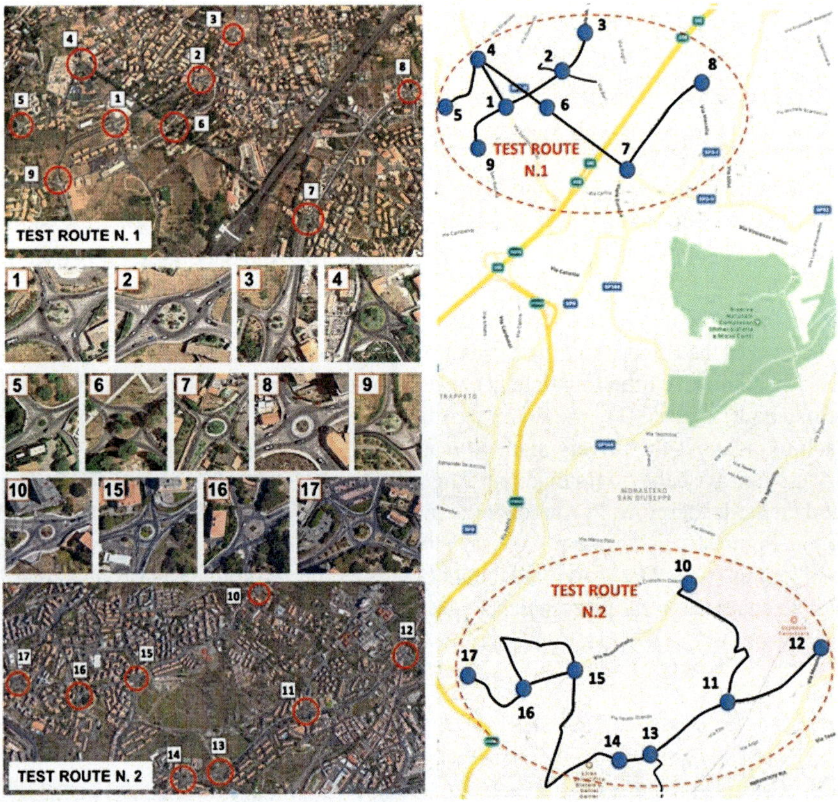

Fig. 10.1 Details of test route 1 and 2 and of the 13 roundabouts considered for the study

behaviours) and unsafe behaviours made while exiting the roundabout (exit unsafe behaviours). Table 10.3 shows the drivers' unsafe behaviours types considered. Table 10.4 shows the number of unsafe behaviours registered during the on-road study.

10.3.5 Analytic Method

In order to analyse the characteristics of the roundabouts and the characteristics of the driver that can lead to unsafe driving behaviour, the CHAID (Chi Square Automatic Interaction Detection) method was used.

CHAID method is a technique employed to discover relationships between a dependent variable and other independent variables, where a statistically significant result identifies their mutual dependence and the relationship between them. In the CHAID algorithm proposed by Kass [33], a chi-squared test (χ^2, chi-square statistic)

Table 10.3 Drivers' unsafe behaviours types at single-lane roundabouts

Entry unsafe behaviours		
1		*High speed of approach:* Approaching the roundabout with a speed higher than the speed limits
2	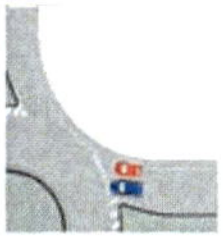	*Parallel entry on entry lane:* Arranging parallel to another vehicle on the entry leg, despite the leg has one lane
3	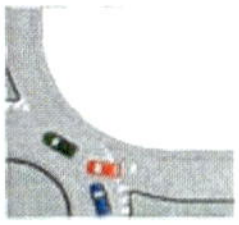	*Selecting unsafe gap:* Selecting unsafe gap when entering the roundabout
4	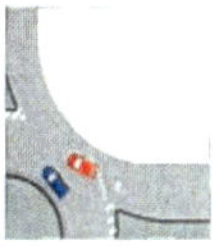	*Parallel entry on circulatory roadway:* Arranging parallel to another vehicle on the circulatory roadway after entering the roundabout, despite the circulatory roadway has one lane
5	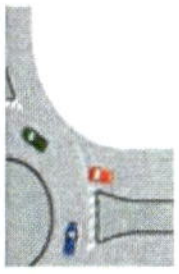	*Rejecting a safe gap:* Rejecting a safe gap when entering the roundabout
Circulation unsafe behaviours		
6		*Parallel circulation:* Arranging parallel to another vehicle on the circulatory roadway when circulating the roundabout, despite the circulatory roadway has one lane
7		*Giving way:* Giving way to incoming vehicle when circulating the roundabout
8		*No circulation:* The vehicle takes the 1st exit without deviation of trajectory
Exit unsafe behaviours		
9	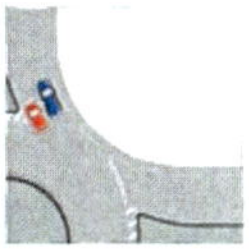	*Parallel exit:* Arranging parallel to another vehicle on the exit leg
10		*Driving over the splitter island:* Driving over the splitter island when exiting the roundabout

(continued)

Table 10.3 (continued)

11	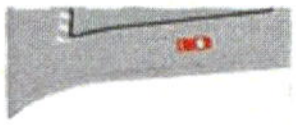	*High exit speed:* Exiting the roundabout with a speed higher than the speed limits

Table 10.4 Number of unsafe behaviours observed

		Number	Percentage [%]
Entry unsafe behaviours			
1	High speed of approach	1862	34.17
2	Parallel entry on entry lane	152	2.79
3	Selecting unsafe gap	759	13.93
4	Parallel entry on circulatory roadway	140	2.57
5	Rejecting a safe gap	282	5.17
Circulation unsafe behaviours			
6	Parallel circulation	455	8.35
7	Giving way	462	8.48
8	No circulation	443	8.13
Exit unsafe behaviours			
9	Parallel exit	226	4.15
10	Driving over the splitter island	576	10.57
11	High exit speed	93	1.71
Total		5450	100.00

is applied to determine the splitting condition. It is mainly used to calculate the degree of dependence between several variables–the larger the value calculated by χ^2, the higher the degree of dependence and the probability value of the variable. Moreover, a probability value is used to determine whether to continue the splitting process in the CHAID algorithm to estimate all the possible predictive variables. In this method, the significance levels of the differences between the various categories of dependent variables are tested for each variable.

Particularly, the CHAID algorithm is used to calculate attribute branches. In the CHAID branching process, each node is branched on the basis of the selected dependent variables, and the chi-squared test is used as the standard for branching. This implies that the branching is conducted whether the classification attribute is significant or not. If the branches have no significant difference, they are merged into the same branch. Conversely, if the branches differ significantly, the branch is retained and the branching process is conducted on the next layer.

CHAID analysis tries to look for patterns in datasets with multiple categorical variables and builds a model in form of a decision tree by splitting the sample or the target dependent variable. CHAID analysis is best for data with large sample size, as the predictor variables are repeatedly split to get categories with equal number of observations to get a final outcome or till CHAID analysis does not find any significantly discriminating in order to receive predictor any more. Since CHAID is best applicable in scenario for categorized value instead of continuous and clearly shows how variables best combine to explain the outcome in a given dependent variable, to outperform better than other statistical tools, such as basic kinds of regression. CHAID is convenient to use in the case of multiple variables as it offers segmentation of one variable based on the effect of combination of a range of independent variables.

In this study, the unsafe driving behaviours recorded during the driving test (as defined in Table 10.3) were fixed as dependent variable. Particularly, entry unsafe behaviours, circulation unsafe behaviours and exit unsafe behaviours were considered separately. The independent variables considered are instead the drivers' characteristics, the roundabout characteristics and the manoeuvres showed in Table 10.5.

10.3.6 Results and Discussions

The CHAID analysis was applied separately for entry, circulation and exit unsafe behaviours. For each type of unsafe behaviour, a tree diagram was obtained.

Figure 10.2 shows the tree diagram obtained for entry unsafe behaviours. The dependent variable considered is entry unsafe behaviour, which can assume values from 1 to 5 (see Table 10.3). The independent variables considered are the ten variables shown in Table 10.4. All entry unsafe behaviours are divided into 22 subgroups from root node to leaf nodes through different branches. Seven variables out of the original set of ten provide a significant explanation of the entry unsafe behaviours. The tree structure involves therefore seven splitting variables: manoeuvre (chi-square = 126.613; *p-value* = 0.000), entry radius (chi-square = 29.183; *p-value* = 0.000; chi-square = 90.515; *p-value* = 0.000), roundabout radius (chi-square = 49.447; *p-value* = 0.000), gender (chi-square = 13.262; *p-value* = 0.010), entry width (chi-square = 58.892; *p-value* = 0.000), splitter island (chi square = 29.719; p-value = 0.000; chi-square = 12.629; *p-value* = 0.013) and circulatory roadway width (chi square = 18.059; p-value = 0.001) meaning that the variable entry radius is responsible for two partitions and the variable splitter island is responsible for two partitions. The manoeuvre is therefore the variable with the most significant effect on entry unsafe behaviours. Among the roundabout characteristics, the entry radius and the roundabout radius are the variables with the most significant effect on entry unsafe behaviours. Among the driver's characteristics, only the gender has a significant effect on entry unsafe behaviours.

From the analysis of Fig. 10.2 it can be observed that the most frequent entry unsafe behaviour is unsafe behaviour 1 (i.e. *High speed of approach*). The percentage of unsafe behaviour 1 at node 0 is 58.3%. Unsafe behaviour 3 (i.e. *Selecting unsafe gap*) is very frequent too. The percentage of unsafe behaviour 3 at node 0 is 23.8%.

Table 10.5 CHAID analysis independent variables

	Variable	Categories
Driver's characteristics	Gender	Male Female
	Age	18–25 26–50 51–65
Roundabout characteristics	Roundabout Radius (RR)	<14 m 14–20 m >20 m
	Circulatory roadway width (CW)	<6 m 6–8 m >8 m
	Entry width (EnW)	<4 m 4–6 m >6 m
	Exit width (ExW)	<4 m 4–6 m >6 m
	Entry radius (EnR)	<10 m 10–30 m 31–50 m >50 m
	Exit radius (ExR)	<10 m 10–30 m 31–50 m >50 m
	Splitter island	Painted Raised
Manoeuvre	Manoeuvre made	1st exit 2nd exit 3rd exit U-turn

The seven splitting variables lead to the tree being divided into three levels. The first optimal split in node 0 is according to manoeuvre, which classifies entry unsafe behaviours into three groups: the tree shows respectively 27.4% of unsafe behaviours for 1st exit manoeuvres, 61.1% of unsafe behaviours 2nd and 3rd exit manoeuvres overall and 11.6% of unsafe behaviours for U-turn manoeuvres.

Unsafe behaviour 1 (i.e. *High speed of approach*) for U-turn manoeuvres is the unsafe behaviour with the highest percentage of the first level of the tree (70%). The distribution of unsafe behaviour 1 is similar for 1st and 2nd exit manoeuvres (respectively 52.7% and 58.5%). On the other hand, it is interesting to note that the distributions of the other unsafe behaviours change according to the exit that the driver has to take. Unsafe behaviour 4 (i.e. *Parallel entry on circulatory roadway*) is significantly major for 1st exit manoeuvres (9.6%) rather than for 2nd and 3rd exit manoeuvres (2.7%) and for U-turn (0.8%). This suggests that the drivers who take

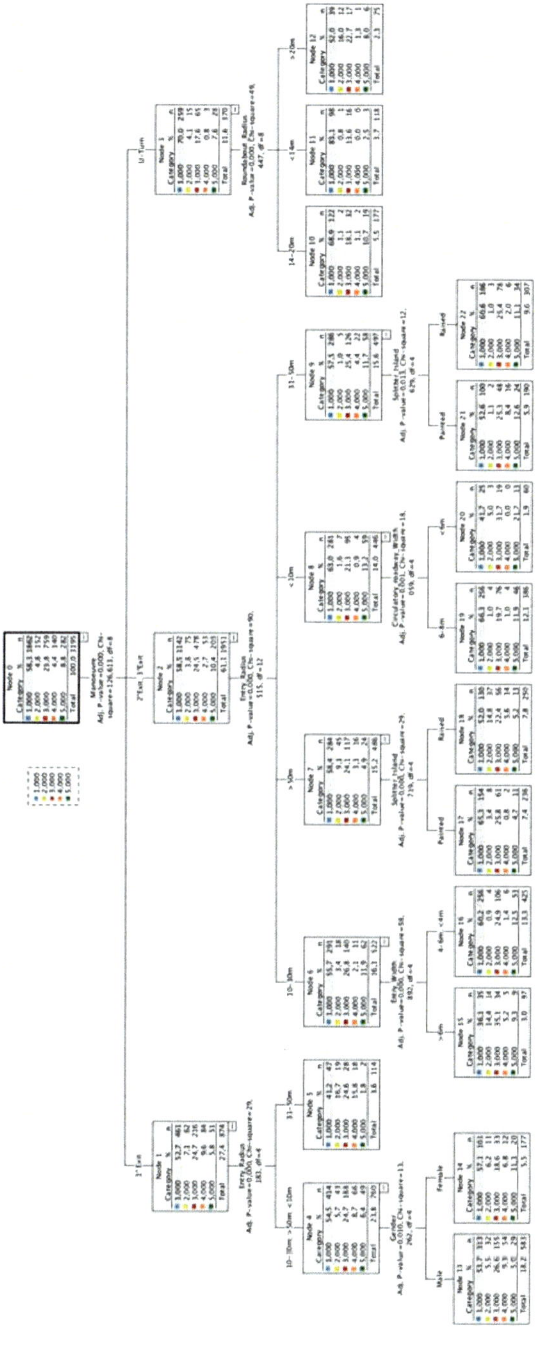

Fig. 10.2 Tree diagram obtained from CHAID analysis for entry unsafe behaviours

the 1st exit are led to believe that they do not interfere with circulating vehicles. This conclusion is supported by the low percentage of unsafe behaviour 5 (i.e. *Rejecting a safe gap*) for 1st exit manoeuvre (5.8%). In the second level of the tree, the group including 1st exit leads to another split based on entry radius; the group including 2nd and 3rd exit leads to another split based on entry radius too; while the group including U-turn leads to another split based on the roundabout radius. It is noteworthy that unsafe behaviour 1 (i.e. *High speed of approach*) is much more evident on roundabout with smaller diameter (respectively 83.1% for RR < 14 m, 68.9% for 14 m ≤ RR ≤ 20 m and 52.0% for RR > 20 m). Conversely, unsafe behaviour 3 (i.e. *Selecting unsafe gap*) and unsafe behaviour 5 (i.e. *Rejecting a safe gap*) are more frequent for roundabouts with bigger diameter. In the third level of the tree, the gender segments the group including 1st exit manoeuvre and EnR > 50 m or 10 m ≤ EnR ≤ 30 m or EnR < 10 m into two subgroups. The leaf nodes 13 and 14 shows that for these conditions, men made more entry unsafe behaviours than women. For the groups 2nd/3rd exit manoeuvre, in case of big entry radius (30 m ≤ EnR ≤ 50 m and EnR > 50 m), the splitter island has a significant effect. While for medium or small entry radius (10 m ≤ EnR ≤ 30 m and EnR < 10 m) the entry width and the circulatory roadway width have, respectively, a significant effect. It can be observed that for 2nd/3rd exit manoeuvre, small Entry Radius (EnR < 10 m) and narrow circulatory roadway (CW < 6 m) the number of entry unsafe behaviours made is the minimum (1.9%).

For 2nd and 3rd exit manoeuvres, when small entry radius (En < 30 m) are paired with small entry width (EnW < 6 m) or with big circulatory roadway width (6 m ≤ CW ≤ 8 m), the number of unsafe behaviour 1 (i.e. *High speed of approach*) is greatly increased. The driver's characteristics does not seem to strongly affect entry unsafe behaviours. However, the gender has a certain influence on unsafe behaviour 3 (i.e. *Selecting unsafe gap*) for 1st exit manoeuvres. It seems indeed that in these cases men are more inclined than women to select unsafe gap (unsafe behaviour 3, respectively 26.6% for men and 18.6% for women). On the other hand, women are more likely to reject a safe gap (unsafe behaviour 5, respectively 11.3% for woman and 5.0% for men).

Figure 10.3 shows the tree diagram obtained for circulation unsafe behaviours. The dependent variable considered is circulation unsafe behaviour, which can assume values from 6 to 8 (see Table 10.3). The independent variables considered are the ten variables shown in Table 10.4. All circulation unsafe behaviours are divided into 16 subgroups from root node to leaf nodes through different branches. Six variables out of the original set of ten provides a significant explanation of the circulation unsafe behaviours. The tree structure involves therefore six splitting variables: manoeuvre (chi-square = 1119.384; *p-value* = 0.000), exit radius (chi-square = 14.683; *p-value* = 0.001), entry width (chi-square = 13.265; *p-value* = 0.001), roundabout radius (chi-square = 11.931; *p-value* = 0.002), gender (chi-square = 7.615; *p-value* = 0.006) and exit width (chi square = 11.025; p-value = 0.003; chi square = 9.041; p-value = 0.008) meaning that the variable exit width is responsible for two partitions. The manoeuvre is therefore the variable with the most significant effect on circulation unsafe behaviours. Among the roundabout characteristics, the exit radius, the entry

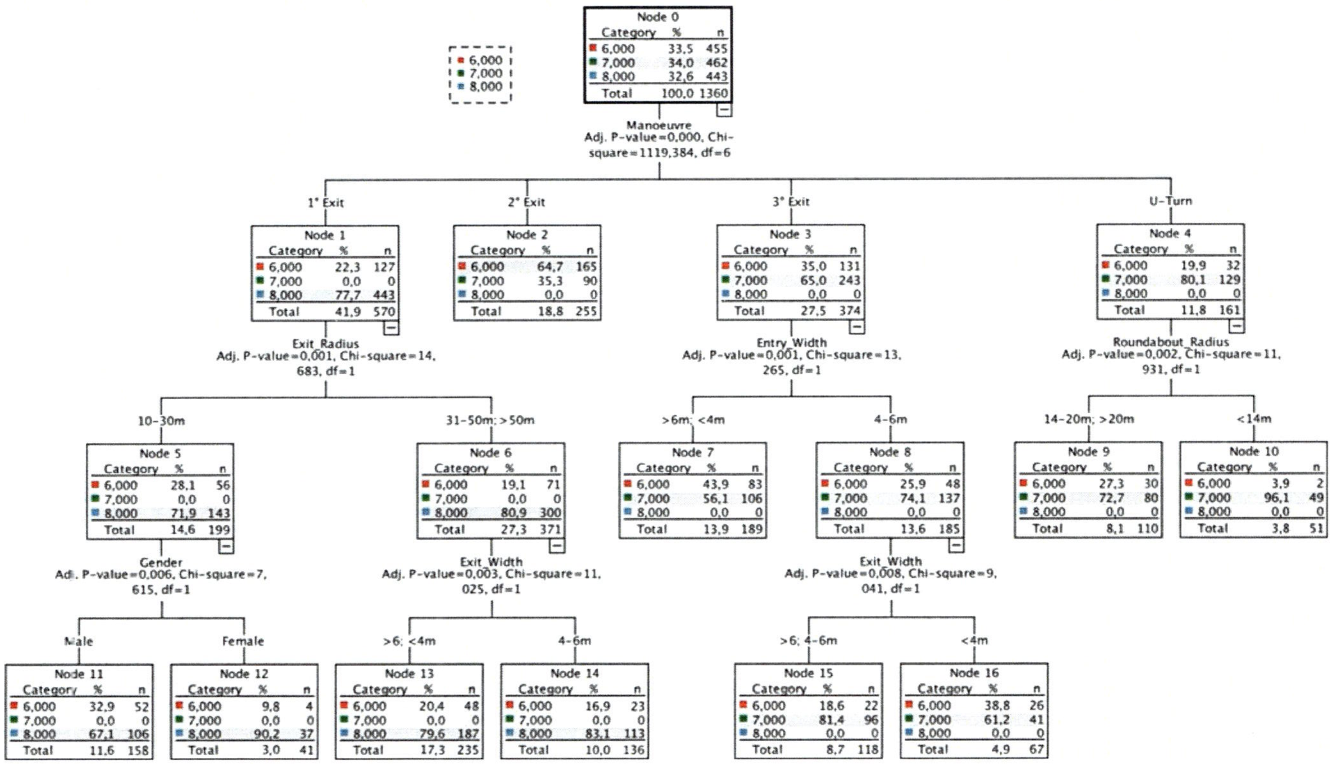

Fig. 10.3 Tree diagram obtained from CHAID analysis for circulation unsafe behaviours

width and the roundabout radius are the variables with the most significant effect on circulation unsafe behaviours. Among the driver's characteristics, only the gender has a significant effect on circulation unsafe behaviours.

From the analysis of Fig. 10.3 it can be observed that there is no a circulation unsafe behaviour more frequent than the others. The percentages of unsafe behaviours 6, 7 and 8 at node 0 are indeed comparable (33.5%, 34.0% and 32.6% respectively).

The six splitting variables lead to the tree being divided into three levels. The first optimal split in node 0 is according to manoeuvre, which classifies circulation unsafe behaviours into four groups: the tree shows respectively 41.9% of unsafe behaviours for 1st exit manoeuvre, 18.8% of unsafe behaviours for 2nd exit manoeuvre, 27.5% of unsafe behaviours for t 3rd exit manoeuvre and 11.8% of unsafe behaviours for U-turn manoeuvre. The majority of circulation unsafe behaviours are therefore made for 1st exit manoeuvres. It has to be noted that unsafe behaviour 8 (i.e. *No circulation*) can be made only for 1st exit manoeuvres. This is the reason why the percentage of unsafe behaviour 8 is 0.0% at nodes 2, 3 and 4. In the same way, unsafe behaviour 7 (i.e. *Giving way*) can be made only for 2nd, 3rd and U-turn manoeuvres. For this reason, the percentage of unsafe behaviour 7 at node 1 is 0.0%. Moreover, it is interesting to observe that the percentage of unsafe behaviour 7 gradually increases for 2nd exit, 3rd exit and U-turn manoeuvres (respectively 35.3%, 65.0% and 80.1%). This leads to the conclusion that the percentage of unsafe behaviour 7 increases with the number of legs that the driver encounters before exiting the roundabout. For 2nd exit manoeuvre unsafe behaviour 6 (i.e. *Parallel circulation*) is the most common (64.7%), while unsafe behaviour 7 (i.e. *Giving way*) is less frequent (35.3%). For 3rd exit manoeuvres the distribution is the opposite, i.e. unsafe behaviour 7 (i.e. *Giving way*) is the most common (65.0%), while unsafe behaviour 6 (i.e. *Parallel circulation*) is less frequent (35.0%). Both manoeuvres are characterized by long trajectories, but during 2nd exit manoeuvres the driver encounters only one leg before exiting the roundabout. Because of these unsafe behaviours 7 are less frequent for 2nd exit manoeuvres. On the other hand, during 3rd exit manoeuvres the driver encounters two legs before exiting the roundabout. This is why unsafe behaviour 7 are more frequent for 3rd exit manoeuvres.

In the second level of the tree, the group including 1st exit leads to another split based on exit radius. It can be observed that the majority of unsafe behaviours is made for big exit radius (i.e. ExR > 30 m). The group including 3rd exit leads to another split based on entry width, even if the percentages of unsafe behaviours obtained for the two categories of entry width are comparable (respectively 13.9% and 13.6%). However, it is noteworthy that the percentage of unsafe behaviour 7 (i.e. *Giving way*) for 4 m $\leq$ EnW $\leq$ 6 m (74.1%) is the highest. The group including U-turn leads to another split based on the roundabout radius. It is interesting to observe that for U-turn manoeuvres the percentage of unsafe behaviour 6 (i.e. *Parallel circulation*) is very small (3.9%) for small radius of roundabout (RR < 14 m), while is definitely higher (27.3%) for bigger radius of roundabout (RR > 14 m). This suggests that the U-Turn manoeuvre on a circulatory roadway with small radius is constrained and makes it very difficult to arrange on parallel lines.

In the third level of the tree, the gender segments the group including 1st exit manoeuvre and ExR > 50 m or 10 m $\leq$ ExR $\leq$ 30 m into two subgroups. Male drivers made more circulation unsafe behaviours than female drivers. Men made 67.1% of unsafe behaviour 8 and 32.9% of unsafe behaviour 6, while women made more unsafe behaviour 8 (90.2%).

The exit width segments into two subgroups both the group including 1st exit manoeuvre and 30 m $\leq$ ExR $\leq$ 50 m and the group including 3rd exit manoeuvre and 4 m $\leq$ EnW $\leq$ m. It is noteworthy the influence of entry width for 3rd exit manoeuvre: the percentage of circulation unsafe behaviours for 4 m $\leq$ EnW $\leq$ 6 m (8.7%) is almost double than the percentage for EnW < 4 m (4.9%).

Figure 10.4 shows the tree diagram obtained for exit unsafe behaviours. The dependent variable considered is exit unsafe behaviour, which can assume values from 9 to 11 (see Table 10.3). The independent variables considered are the ten variables shown in Table 10.4. All exit unsafe behaviours are divided into 13 subgroups from root node to leaf nodes through different branches. Four variables out of the original set of ten provides a significant explanation of the exit unsafe behaviours. The tree structure involves therefore four splitting variables: manoeuvre (chi-square = 114.558; *p-value* = 0.000); exit width (chi-square = 58.953; *p-value* = 0.000), splitter island (chi-square = 27.320; *p-value* = 0.000, chi-square = 45.671; *p-value* = 0.000) and exit radius (chi-square = 10.960; *p-value* = 0.013) meaning that the variable splitter island is responsible for two partitions. The manoeuvre is therefore the variable with the most significant effect on exit unsafe behaviours. Among the driver's characteristics, no variables have a significant effect on exit unsafe behaviours.

From the analysis of Fig. 10.4 it can be observed that unsafe behaviour 11 (i.e. *High exit speed*) is the most frequent. The percentage of unsafe behaviour 11 at node 0 is indeed 64.4%. This is a confirmation of the results obtained for entry unsafe behaviour. Drivers are indeed led to try to make the manoeuvres on roundabouts as fast as possible. This leads to aggressive behaviours, which are evident on exit unsafe behaviours too. 65% of exit unsafe behaviours regards indeed *High exit speed* (i.e. unsafe behaviour 11) and about 25% regards *Driving over the splitter island* (i.e. unsafe behaviours 10).

The four splitting variables lead to the tree being divided into three levels. The first optimal split in node 0 is according to manoeuvre, which classifies exit unsafe behaviours into four groups: the tree shows 34.6% of unsafe behaviours for 2nd exit manoeuvre, 24.6% of unsafe behaviours for 1st exit manoeuvre, 26.1% of unsafe behaviours for 3rd exit manoeuvre and 14.6% U-turn. The majority of exit unsafe behaviours are therefore made for 2nd exit manoeuvre. It is interesting to observe that the percentage of unsafe behaviour 11 (i.e. *High exit speed*) for U-turn manoeuvre is the highest (93.9%). For 1st and 3rd exit manoeuvres unsafe behaviour 11, despite being prevalent, it is comparable to other unsafe behaviours. About 40% of unsafe behaviours for 3rd exit manoeuvres regards *Driving over the splitter island* (i.e. unsafe behaviours 10). For 1st exit manoeuvres the percentage of unsafe behaviour 11 is particularly high (28.6%) and the percentage of unsafe behaviour 9 (i.e. *Parallel exit*) is relevant too (21.4%). The results suggest therefore that when drivers take the 1st exit they tend to make a direct and fast manoeuvre without significant changes

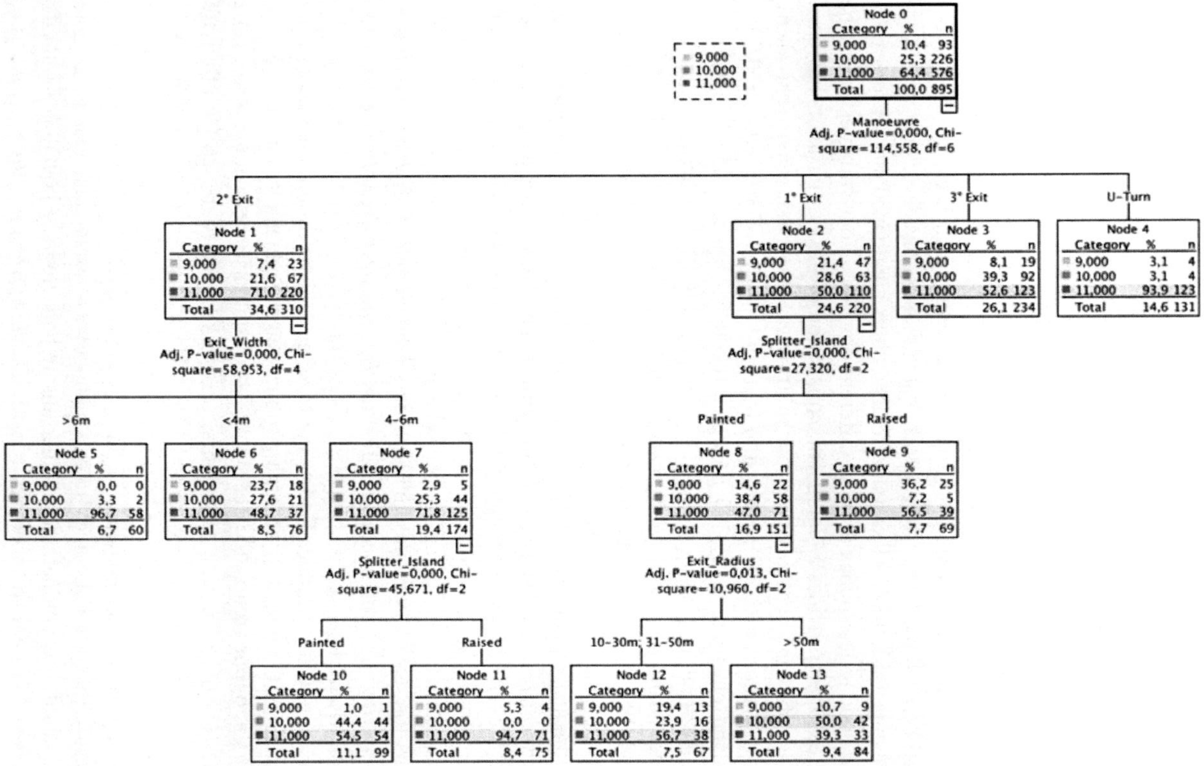

Fig. 10.4 Tree diagram obtained from CHAID analysis for exit unsafe behaviours

in the turning radius; this often results in driving over the splitter islands and in arranging parallel to other vehicles on the exit leg. On the other hand, when the drivers take the 3rd exit the trajectories lead much more rarely to arrange parallel to other vehicles on the exit leg, even if they often lead to drive over the splitter islands (principally for painted island).

In the second level of the tree the exit width segments into two subgroups the group including 2nd exit manoeuvre. In case of ExW > 6 m the percentage of unsafe behaviour 11 (i.e. *High exit speed*) is very high (96.7%), while it is smaller for ExW < 4 m (48.7%). The group including 1st exit manoeuvre leads to another split based on splitter island. It can be observed that the percentage of exit unsafe behaviours is major for painted splitter islands (16.9%) rather than for raised splitter island (7.7%). In the third level of the tree, the splitter island segments the group including 2nd exit manoeuvre and 4 m ≤ ExW ≤ 6 m into two subgroups. The exit radius segments the group including 1st exit manoeuvre and painted splitter island into two subgroups.

10.4 Conclusions

Proper road design is crucial to prevent unsafe driving behaviours in traffic and less unsafe behaviours would result in fewer accidents. This chapter aimed to identify the contributing factors affecting the drivers' unsafe behaviours at single-lane roundabouts. Unsafe behaviours are evaluated based on the behaviour of sixty-six drivers who performed an on-road test around the suburbs surrounding the city of Catania. Three CHAID analysis were developed in order to analyse the influence of driver's characteristics, roundabout characteristics and manoeuvre on entry, circulation and exit unsafe behaviours respectively.

The results of this study show that the manoeuver that the driver have to perform in the roundabout is the variable that most influences unsafe driving behaviours, both when entering, circulating and exiting. The unsafe behaviours observed more frequently are the high speed with which the drivers travel the roundabout and the lack of deflection in the execution of the right turn manoeuver (1st Exit).

Furthermore, the study found that the roundabout characteristics (lanes width, radii, splitter islands) influence unsafe driving behaviours at roundabouts. In particular: (1) the entry radius and the roundabout radius are the variables with the most significant effect on entry unsafe behaviours; (2) the majority of circulation unsafe behaviours are made for big exit radius (over 30 meters); (3) the exit width is the geometric variable with the most significant effect on exit unsafe behaviours, mainly causing high exit speed especially in the execution of the crossing manoeuvre. On the other hand, among the driver's characteristics (gender, age and mean of transport) only the gender was found to be significant (with low significance).

The results of this study contribute to increase knowledge about the relationship between road design and human behaviour in order to encourage the design of the physical road layout in such a way that incorrect driving behaviour is not viable in physical terms. The main message is that the human factor both in terms of behaviour and physical vulnerability is the main problem in transport safety. Therefore, it is necessary to optimize the design criteria and, at the same time, to take a series of technical and organizational actions to neutralise the human factor to achieve safety of the transportation infrastructures.

These findings provide the foundation for further research and will be used in further research to develop improved models that explicitly account for the different unsafe types of behaviour in other roundabout configurations. To expand the data presented here, future work should also seek to address the nature of similar types of behaviour occurring at multilane roundabouts and turbo roundabouts. Precisely the turbo roundabouts, through the organization of the lanes of the circulatory roadway that forces the vehicles to move within obligatory paths, could represent the tool to heavily influence the behaviours of the drivers, to the benefit of road safety performance.

References

1. Fastenmeier W, Gstalter H (2007) Driving task analysis as a tool in traffic safety research and practice. Safety Sci 952–979
2. Bastos Silva A, Santosa S, Vasconcelos L, Seco A, Silva JP (2014) Driver behavior characterization in roundabout crossings. In: 17th Meeting of the EURO Working Group on Transportation, EWGT2014, 2–4 July 2014, Sevilla, Spain. Transportation Research Procedia 3, 80 – 89. https://doi.org/10.1016/j.trpro.2014.10.093
3. Zhao M, Kathner D, Jipp M, Soffker D, Lemmer K (2017) Modeling driver behavior at roundabouts: results from a field study. In: IEEE Intelligent Vehicles Symposium (IV). https://doi.org/10.1109/ivs.2017.7995831
4. St-Aubin P, Saunier N, Miranda-Moreno L, Ismail K (2013) Use of computer vision data for detailed driver behavior analysis and trajectory interpretation at roundabouts. Transp Res Rec B 2389:65–77
5. Mudgal A, Hallmark S, Carriquiry A, Gkritza K (2014) Driving behavior at a roundabout: a hierarchical bayesian regression analysis. Transp Res D-Tr E 26:20–26
6. Distefano N, Leonardi S, Pulvirenti G, Romano R, Merat N, Boer E, Woolridge E (2020) Physiological and driving behaviour changes associated to different road intersections. Eur Transp Trasporti Europei 77
7. Alshannaq M, Imam R (2020) Evaluating of safety performance of roundabout. Transp. Probl. 11(1):141–152. https://doi.org/10.21307/tp-2020-013
8. Abojaradeh M, Jrew B, Al-Ababsah H, Al-Talafeeh A (2014) The effect of driver behavior mistakes on traffic safety. Civil Environ Res 6(1):39–54
9. Uzondu C, Jamson S, Lai F (2019) Investigating unsafe behaviours in traffic conflict situations: an observational study in Nigeria. J Traffic Transp Eng (English Edition) 6(5):482–492. https://doi.org/10.1016/j.jtte.2018.06.00
10. Lee JD, Young KL, Regan MA (2009) Defining driver distraction. In: Regan MA, Lee JD, Young KL (eds), Driver distraction: theory, effects, and mitigation. CRC Press, Boca Raton, pp. 31–40
11. Dabbour E, Al Awadhi M, Aljarah M, Mansoura M, Haider M (2018) Evaluating safety effectiveness of roundabouts in Abu Dhabi. IATSS Res 42(4):274–8. https://doi.org/10.1016/j.iatssr.2018.04.003
12. Małecki K, Wątróbski J, Wolski W (2017) A cellular automaton based system for traffic analyses on the roundabout. In: Conference on computational collective intelligence technologies and applications. Springer, Cham, pp 56–65

13. Sacchi E, Bassani M, Persaud B (2011) Comparison of safety performance models for urban roundabouts in Italy and other countries. Transp Res Rec 2265:253–9. https://doi.org/10.3141/2265-28
14. Kamla J, Parry T, Dawson A (2016) Roundabout accident prediction model: random-parameter negative binomial approach. Transp Res Rec 2585:11–19. https://doi.org/10.3141/2585-02
15. Mandavilli S, McCartt A, Retting RA (2009) Crash patterns and potential engineering countermeasures at Maryland roundabouts. Traffic Inj Prev 10(1):44–50
16. Montella A (2011) Identifying crash contributory factors at urban roundabouts and using association rules to explore their relationships to different crash types. Accident Anal Prev 43(4):1451–1463
17. Rodegerdts L, Blogg M, Wemple E, Myers E, Kyte M, Dixon M, Carter D (2007) Roundabouts in the United States (National Cooperative Highway Research Program Report No. 572). Transportation Research Board, Washington, DC
18. Ramisetty-Mikler S, Almakadma A (2016) Attitudes and behaviors towards risky driving among adolescents in Saudi Arabia. Int J Pediatr Adolesc Med. 3(2):55–63
19. Bener A, Yildirim E, Bolat E, Özkan T, Lajunen T (2016) The driver behaviour questionnaire as an accident predictor in cross-cultural countries in Qatar and Turkey: Global Public Health Problem. Brit J Med Res 15(7):1–9
20. Al-Rukaibi F, Ali MA, Aljassar A, Al-Abdulmuhsen EL (2008) A study of driver behaviorwith regard to traffic control devices. In: Efficient transportation and pavement systems: characterization, mechanisms, simulation and modeling, p 73
21. Al-Saleh K, Bendak S (2012) Drivers'behaviour at roundabouts in Riyadh. Int J Inj ControlSaf Promot 19(1):19–25
22. Distefano N, Leonardi S, Consoli F (2019) Drivers' preferences for road roundabouts: a study based on stated preference survey in Italy. KSCE J Civ Eng 23(11):4864–4874. https://doi.org/10.1007/s12205-019-1363-9
23. Larsson P, Dekker SWA, Tingvall C (2010) The need for a systems theory approach to road safety. Saf Sci 48(9):1167–1174
24. Young KL, Salmon PM, Lenné MG (2013) At the cross-roads: An on-road examination of driving errors at intersections. Accident Anal Prev 58:226–234. https://doi.org/10.1016/j.aap.2012.09.014
25. Jameel AK, Evdorides H (2020) Developing a safer road user behaviour index. IATSS Res, Available online 10 July 2020. In Press. https://doi.org/10.1016/j.iatssr.2020.06.006
26. Reason J, Manstead ASR, Straling SG, Baxter JS, Campbell K (1990) Errors and violations on the road: a real distinction? Ergonomics 33:1315–1332. https://doi.org/10.1080/00140139008925335
27. Salmon PM, Regan M, Johnston I (2006) Human error and road transport: phase one- literature review. Monash University Accident Research Centre Report
28. Mandiartha P, Al Jahdhamia M, Badr A, Hayne G, Duffield C, Thompson R (2020) Effect of roundabout design on the behaviour of road users: a case study of roundabouts in Oman. In: Proceedings of 8th Transport Research Arena TRA 2020, April 27–30, Helsinki, Finland
29. Papantoniou P, Yannis G, Christofa E (2019) Which factors lead to driving errors? A structural equation model analysis through a driving simulator experiment. IATSS Res 43:44–50
30. McKnight GA, Khattak AJ, Bishu R (2008) Driver characteristics associated with knowledge of correct roundabout negotiation. Transp Res Rec 2078(1):96–99
31. Mensah S, Eshragh S, Faghri A (2010) A critical gap analysis for modern roundabouts. In: Transportation research board 89th annual meeting, Washington, DC
32. Belz NP, Aultman-Hall L, Gårder PE, Lee BHY (2014) Event-based framework for noncompliant driver behavior at single-lane roundabouts. Transp Res Rec 2402(1):38–46
33. Kass GV (1980) An exploratory technique for investigating large quantities of categorical data. Appl Stat-J Roy St C 29(2):119–127. https://doi.org/10.2307/2986296

Chapter 11
Naturalistic Driving Study: Methodological Aspects and Exemplary Analysis of a Long Roadwork Zone

Anton Pashkevich, Jacek Bartusiak, Tomasz E. Burghardt, and Matúš Šucha

11.1 Introduction

During the past 20 years, 88,078 people have been killed and 1,061,267 wounded in 852,711 road accidents in Poland [1]. The associated financial losses can be estimated, through an extrapolation of an analysis from road safety advocates [2], at 889 $\times 10^9$ Polish złoty (PLN), which according to official exchange rates would be 219 $\times 10^9$ euro (EUR) or 267 $\times 10^9$ United States dollars (USD). According to police statistics from 2019, 20.8% of vehicular accidents in Poland could be attributed to excessive speed [1]. Given enormous death and injury toll and financial losses that, combined for the two decades approached 39% of Poland's Gross Domestic Product in 2019, identification of factors contributing to such poor safety record is of utmost importance.

While there are many methods for analysis of road users' behaviour and assessment of road infrastructure (e.g. [3–6]), naturalistic driving studies are surprisingly rare. While naturalistic evaluations lack the convenience of simulating a plethora of

A. Pashkevich · J. Bartusiak
Politechnika Krakowska, Ul. Warszawska 24, 31-155 Kraków, Poland

A. Pashkevich (✉) · M. Šucha
Palacký University Olomouc, Křížkovského 511/8, 771 47 Olomouc, Czech Republic
e-mail: apashkevich@pk.edu.pl

M. Šucha
e-mail: matus.sucha@upol.cz

T. E. Burghardt (✉)
M. Swarovski GmbH, Industriestraße 10, 3300 Amstetten, Austria
e-mail: tomasz.burghardt@swarco.com

E. Macioszek and G. Sierpiński (eds.), *Research Methods in Modern Urban Transportation Systems and Networks*, Lecture Notes in Networks and Systems 207, https://doi.org/10.1007/978-3-030-71708-7_11

factors that can be easily set under laboratory conditions, they are capable of identifying aspects and distractions that are frequently missed, overestimated, or underrepresented during simulator experiments [7]. In Poland, naturalistic driving studies were done within the scope of UDRIVE project [8], but no follow-up experiments were done.

While initially the work plan was to concentrate on overtaking drivers' speeds, it was quickly realised that frequently the speed limit was not obvious. Hence, the analysis was expanded to incorporate assessment of road signage. Based on the results, it is being argued that while Driver Behaviour Questionnaire (DBQ) separates events into violations and errors [9], at the analysed road stretch numerous violations could be classified as errors because of inappropriate signage and the absence of self-explaining infrastructure. The outcome from experiments run according to the presented methodology could be utilised to verify whether road accidents are indeed caused mostly by driver inattention [10, 11], or whether the infrastructural deficiencies could play equally critical role [12, 13].

Whereas road safety audits can provide a comprehensive and quite exhaustive information and pinpoint the deficiencies [14], they could be time-consuming and occasionally overweighing some factors while missing simple solutions. In addition, safety audit requires the presence of the analysing crew on-site, which might affect behaviour of the drivers, or obstruct the traffic. Analysis of recording from a readily available dashboard camcorder was seen as an inexpensive, simple, and not obstructive methodology, which was, to the best of our knowledge, never so far utilised for assessment of road features or drivers' behaviour. Thus, proposed is new and expedient methodology to evaluate road features and speeds of vehicles from the perspective of a driver, while remaining unbiased. The drawback of evaluating the road from just one perspective appears marginal in comparison with the ease of such assessment, but it requires an acknowledgement. While expansion of this effort and application to other similar situations is undergoing, the newly used methodology is described herein and a plethora of results are given. To the best of our knowledge, this is the first assessment of a road construction zone from naturalistic driving data.

11.2 Experimental

11.2.1 Naturalistic Driving: Driver and Equipment

For the study, the test driver was a native of Poland, holding class B licence (at the time of issue, it was a "professional class B licence", which required psychotechnical exam and allowed for driving of government-owned vehicles; that licence subclass was abandoned in 1990), who had travelled approximately 1.8 million kilometres during over 30 years of driving experience, for business and leisure, during residences in Poland, Austria, Germany, the United States, and Romania, but also with major driving experience in Czech Republic, Croatia, Slovakia, France, and Argentina.

The test vehicle was an estate car, kerb weight 1597 kg, equipped with 1968 cm^3 Diesel engine (Euro 6 emission class) capable of providing 110 kW power at 2375 min^{-1} and direct-shift gearbox transmission. As much as possible, driving was done with the utilisation of Adaptive Cruise Control (ACC) feature, which maintained not only constant speed (usually within ± 1 km/h of the set value), but also assured a safe distance from preceding vehicle. The vehicle was driven with obedience of rules of the road—to the extent that it was possible and reasonable—with particular attention to speed limit.

Data collection was done through a dual dashboard camcorder that was recording video at 30 frames per second, with resolution 1920 × 1080 pixels, at field of view 140°. The companion synchronised camcorder had the same parameters. The camcorder was equipped with a receiver of Global Positioning System (GPS) data, at frequency 1 Hz, from which speed was obtained. The GPS information was recorded separately as NMEA file format developed by National Marine Electronics Association, from which it was extracted for analyses.

All of the ethical guidelines set by the participating academic institutions were followed throughout data collection and processing; only anonymous averages are provided and all reasonable care was taken to prohibit disclosure of any data that could cause privacy violations. There is no law that would prohibit such recordings in Poland or in Czech Republic as long as no identifiable information related to the observed vehicles or people is published. All of the images presented herein were taken from the camcorder.

11.2.2 Selection of Roads

The assessed road section is a 65 km long road work zone associated with a complete reconstruction and an upgrade of the main North-South highway in Poland, national route 1, from a dual carriageway unlimited-access 2 + 2 road to a 3 + 3 motorway. In 2015, the road at the analysed stretch was carrying, on average at different measurement sections, Annually Averaged Daily Traffic (AADT) of 33,880 vehicles, out of which 10,546 (31%) were classified as heavy [15]. During the construction, it is a 2 + 1 road, with occasional narrowing to 1 + 1 and seldom switches in direction of the two-lane stretches, which causes some of the traffic to use a parallel road. Generally, the work zone speed limit is 60 km/h, with occasional reductions to 40 km/h and a few short stretches with other speed limits.

Because in few cases the traffic ahead was stopped or moving too slowly, the test driver departed from the analysed road, so the outcome from various drives on various sections, some of them being repeat drives is presented herein. Assessed were only two-lane road sections (with the occasional narrowing to one-lane) between the future interchanges Częstochowa Północ and Piotrków Trybunalski Południe.

For a minor comparison, similar analysis was done in Czech Republic, at four road work zones at the main East-West motorway D1, between Brno and Humpolec, totalling 43.8 km; speed limit at the two-lane stretches was 80 km/h.

11.3 Methodology

11.3.1 Data Processing

Data processing was done with academic version of a video-editing software. For analysis, recorded were make and model of overtaking vehicles as well as violations of rules of the road. In addition, visual assessment of road signs was done. Correctness of horizontal road markings was noted, too.

Speed of overtaking vehicles was obtained from a frame-by-frame analysis: knowing the length of the overtaking vehicle and/or its wheelbase, it was possible to calculate its speed in relation to the test vehicle, which was moving at the velocity obtained from the NMEA file. The analysis was done for both the moment of entering and leaving the camcorders' field of view by the analysed overtaking vehicle. Correctness of such approach was verified at another road stretches through measurements done with a companion vehicle. Additional random verification was done by measuring the time it took the overtaking vehicle to pass a known distance, such as two long stripes of horizontal road marking (i.e. the distance of 24 m) or kilometre posts; correctness of speed from NMEA file was verified similarly. The agreement between thus obtained speeds was within 5%; errors were unavoidably measured when acceleration or deceleration was taking place, which did not change the overall picture. Additional source of errors could occur if the wheelbase or vehicle length were uncertain.

Speed of the overtaking vehicles was calculated according to the Eq. 11.1, where v_{test} is the speed of the test vehicle [km/h], L is the length (or wheelbase) of the overtaking or overtaken vehicle [m], t_1 and t_2 are the times [s] that it took the analysed vehicle to pass through the limit of camcorder's field of view, and 3.6 is the conversion factor to obtain speed in km/h.

$$v_{overtaking} = v_{test} + \frac{2 \cdot L}{t_1 + t_2} \cdot 3.6 \tag{11.1}$$

Speed of the test vehicle was also evaluated and compared with the posted speed limits. In some cases, it was necessary to double-check, through rewinding and confirming road signs, what was the posted speed limit. The difficulty was caused by the rules of the road specifying in Poland that an intersection (but not a driveway) cancels the limitations; however, one very frequently wonders: "was that a driveway or an intersection". In several cases, the doubts remained even after rewinding the recording several times.

To account for speedometer and measurement errors, speeding was considered to occur only after applying an adjustment, calculated per Eq. 11.2, where $v_{adjusted}$ is the adjusted speed limit [km/h], and v_{posted} is the posted speed limit [km/h].

$$v_{adjusted} = \left([(v)]_{posted} \cdot 1.05\right) + 3 \tag{11.2}$$

Table 11.1 Subjective assessment scale for road signs

Rating	Correctness	Visibility	Quality of sign face
3	Appropriate	Good, no obstructions	Good, clearly legible
2	Counterintuitive, information unclear, unnecessary, redundant	Poorly positioned, possibility of not noticing, crooked or spaced inappropriately	Adequate, not affecting overall visibility or legibility, but road sign requires replacement
1	Erroneous or inappropriate, confusing	Positioned incorrectly, could become hard to notice, confirmation sign on left side positioned improperly	Damaged, poor quality, faded
0	Missing road sign when required	Obscured or invisible	Missing road sign when required

The identified vehicles were assigned into several categories to check whether there would be any correlation with speed and other violations of traffic regulations [16–18].

All distance measurements were obtained from calculations based on GPS coordinates obtained from NMEA file; minor errors, particularly associated with the lack of continuity of GPS location recording, must be acknowledged.

11.3.2 Road Signs

Very important and separate part of the analysis included visual assessment of road signs. Vertical road signs were catalogued and evaluated in terms of their correctness, visibility to drivers, and quality. Correctness included also cross-checking for coherence with other signs and road features as well as signalisation provided by horizontal road markings. A subjective 0–3 scale, shown in Table 11.1, was used. If a road sign contained a pendant indicative tablet (a complimentary sign), it was assessed as a whole, but the presence of the tablet was indicated.

11.3.3 Road Markings

Horizontal road markings were evaluated in the course of this effort only for clarity and correctness. Noted were major inadequacies that could affect road safety and cases where the horizontal signalisation was wrong, forcing all of the drivers to break rules of the road. Clarity and legibility of road markings is consistently reported as critical for appropriate lane recognition by machine vision, used for the emerging technology of automated vehicles [19]. The possibility of utilising the naturalistic driving data from dashboard camcorder for estimation of road marking quality and legibility for machine vision shall be explored separately.

11.4 Results

11.4.1 Test Vehicle Speed

Speed of the test vehicle was maintained, as much as possible, between 1 and 5 km/h over the posted speed limit; measured according to GPS data. This minor speeding was done to avoid any claims, no matter how inappropriate and ridiculous, that the test vehicle was blocking the road by driving too slowly and thus prompting other drivers to overtake it. Some mistakes and inadequacies in posted speed limits resulted in driving too fast, which violations in several cases could be classified as errors according to DBQ.

Driving through the analysed stretches totalling 152 km took 2 h 36 min, giving average speed 58 km/h. Driving faster than the posted adjusted speed limit occurred at a distance of 2.7 km (time of 204 s), with an average speeding of 4 km/h. Maximum speeding was driving 64 km/h in a 40 km/h zone at a distance of 443 m that took 25 s. The posted speed limits show in the analysed stretches 13.7 km (9%) was in 40 km/h zone, 133.9 km (88%) in 60 km/h zone, and 4.7 km (3%) had other speed limit. Speed limit from the standard of 60 km/h was changed 38 times, 32 times (average 21 times per 100 km) lowered to 40 km/h and 6 times (average 4 times per 100 km) to other speed. The theoretical average speed should be 56 km/h (4% lower than actual) and travelling the analysed distance should have taken 2 h 40′ (4′ or 3% more than actual).

11.4.2 Overtaking

At the analysed stretch, the test driver overtook only 1 vehicle (a lorry moving at 59 km/h was overtaken at the speed 61 km/h) and was overtaken 374 times. Thus, there were 2.46 overtaking events per 1 km of travel (144 per 1 h of driving or 2.46 per 1 km). The collected data related to overtaking vehicles, separated into various classes, and their speeding (adjusted speed limit was applied) is given in Table 11.2.

Amongst the results one should note that only 3 vehicles overtook the testing vehicle without exceeding the adjusted speed limit. While the average overtaking vehicle was moving at 97 km/h, the fastest one was travelling at 162 km/h (i.e. 102 km/h or 170% faster than the posted speed limit). As highly troublesome is seen very significant speeding by minibuses, with average speed of 98 km/h (i.e. excess of 38 km/h or 63% over speed limit); no coaches or buses were seen at the test stretch. Overtaking with vehicles pulling trailers occurred 3 times—their average speed was 86 km/h and maximum was 96 km/h. Despite nice summer weather, only two motorcycles were encountered—they overtook at speeds of 84 km/h and 101 km/h. Quite worrisome was also overtaking by lorries and articulated lorries, which are not permitted to do this at the test stretch: it occurred 10 times, at speeds 75–89 km/h (average speed 82 km/h) in 60 km/h zone.

Table 11.2 Speeds of overtaking vehicles[a]

		All	Cars	Sport-Utility Vehicles	Minivans and vans	Delivery vehicles < 3.5 t	Minibuses
Number[a]	n	374	242 (65%)	74 (20%)	19 (5%)	8 (2%)	16 (4%)
Average speed [km/h]	$\bar{v}$	97	98	98	95	86	98
Maximum speed [km/h]	v_{max}	162	162	149	118	91	123
Not speeding	n	3 (1%)	3 (1%)	–	–	–	–
	$\bar{v}$	63	63	–	–	–	–
Speeding 1–10 km/h	n	5 (2%)	2 (1%)	2 (3%)	–	–	–
	$\bar{v}$	73	73	73	–	–	–
Speeding 11–20 km/h	n	60 (16%)	32 (13%)	9 (12%)	4 (21%)	4 (50%)	4 (25%)
	$\bar{v}$	82	83	83	81	82	85
Speeding 21–30 km/h	n	102 (27%)	68 (28%)	20 (27%)	6 (32%)	4 (50%)	1 (6%)
	$\bar{v}$	90	90	89	94	91	90
Speeding 31–40 km/h	n	116 (31%)	73 (30%)	28 (37%)	7 (37%)	–	7 (44%)
	$\bar{v}$	99	99	100	100	–	98
Speeding 41–50 km/h	n	43 (11%)	34 (14%)	8 (10%)	–	–	1 (6%)
	$\bar{v}$	110	110	111	–	–	110
Speeding 51–60 km/h	n	29 (8%)	21 (9%)	5 (7%)	1 (5%)	–	2 (13%)
	$\bar{v}$	119	119	120	118	–	122
Speeding 61–70 km/h	n	12 (3%)	8 (3%)	2 (3%)	1 (5%)	–	1 (6%)
	$\bar{v}$	127	131	129	106	–	113
Speeding >70 km/h	n	4 (1%)	3 (1%)	1 (1%)	–	–	–
	$\bar{v}$	152	152	149	–	–	–

[a]Values do not add to 100% because motorcycles and lorries are not included in the table

There were 21 (5.6%) vehicles with foreign number plates. Amongst them, fastest was travelling 123 km/h (speeding 63 km/h). This could be treated as a confirmation of findings that drivers adapt to the local driving style instead of maintaining habits from their home countries [20]. On the other hand, it is equally likely that these drivers were Poles living abroad.

Amongst major additional violations besides speeding, one must list the 10 no-passing violations by articulated lorries and one of the lorries towing a semi-trailer without lights and without number plates (sic!). There were 41 (11.0%) vehicles that changed lanes without using blinkers and 31 (8.3%) used blinkers improperly. In 50 (13.4%) cases, drivers failed to return to the right lane after finishing the overtaking manoeuvre and not overtaking another vehicle.

11.4.3 Road Signs

The number of road signs in Poland is gargantuan, which not only creates a visual clutter, but also makes their proper recognition impossible [21]. The same issue was observed during the test drives. The number of road signs, divided into categories, is given in Table 11.3. Because numerous road signs are necessary due to intersections, their numbers were recalculated per 1 km of road and per intersection. At the test stretch there were counted 2,022 road signs, including 422 (20.9%) that had a complimentary pendant tablets, either symbolic or written and 572 (28.3%) that were repeats on the left side of the road (16 signs were located only on the left side, out of which 12 (75%) were subjectively rated as 0-2). On average, there were 13.28 road signs per 1 km of road, which at the driving speed of the test vehicle was giving 12.96 signs per minute (i.e. a road sign every 4.6 s). It should be noted that in some cases up to 3 road signs were placed together, so quite important becomes the number of posts, also provided in Table 11.3.

The largest concentration of road signs was 14 signs (including 3 with complimentary tablets) at a distance of 85 m. For drivers travelling at the speed limit this would mean 5 road signs per second, which, after comparing with a typical visual fixation time at a road sign of 0.154 s [22], means that driver must spent 77% of driving time to recognize all road signs (without accounting additional recognition or reading time that may be needed for recognition of the pendant tablets). Taking into account that signs attract only 15–20% of driver's attention [21], this task starts to be unrealistic. An example of such road signs concentration is shown in Fig. 11.1.

Assessment of road signs correctness, visibility, and quality is given in Table 11.4. The quality of road signs was acceptable: only 114 (5.6%) of the signs were considered to be in poor condition (ratings 0–2). Visibility (i.e. positioning) were satisfactory in majority of cases, with 70 (3.5%) considered as inadequate (ratings 0–2). Definitely, by far the poorest was correctness: 644 (31.8%) of road signs were rated as incorrect; in addition, the authors subjectively consider 728 (36.0%) as unnecessary and 114 (5.6%) were noted as providing false information.

Table 11.3 Vertical road signs

	All	Warning	Prohibitory[a]	Speed limit	Mandatory	Information	Direction	Unofficial[a]	Other
Number of road signs	2,022	421 (20.8%)	801 (39.6%)	441 (21.8%)	132 (6.5%)	57 (2.8%)	42 (2.1%)	80 (4.0%)	489 (24.2%)
Number of posts	1,655	376 (22.7%)	486 (29.4%)	441 (26.6%)	132 (8.0%)	57 (3.4%)	35 (2.1%)	80 (4.8%)	489 (29.6%)
Number of road signs on the left side (all)	588	101 (17.2%)	281 (47.8%)	152 (25.9%)	20 (3.4%)	12 (2.0%)	2 (0.3%)	21 (3.6%)	151 (25.7%)
Number of road signs on the left side (repeats from right side)	572	98 (23.3%)	275 (34.3%)	149 (33.8%)	16 (12.1%)	12 (21.1%)	2 (4.8%)	21 (26.3%)	148 (30.3%)
Number of road signs on the left side (absent on the right side)	16	3 (19%)	6 (37%)	3 (19%)	4 (25%)	–	–	–	3 (19%)
Number of posts on the left side	445	86 (19.3%)	153 (34.4%)	152 (34.2%)	20 (4.5%)	12 (2.7%)	2 (0.5%)	21 (4.7%)	151 (33.9%)
Supplementary complimentary plaques	470	253 (53.8%)	59 (12.5%)	31 (6.6%)	115 (24.5%)	0 (0%)	0 (0%)	38 (8.1%)	5 (1.1%)
Number of road signs per 1 km	13.28	2.76	1.85	2.90	0.87	0.37	0.28	0.53	3.21
Number of road signs per intersection[(c)]	18.05	3.76	2.51	3.94	1.18	0.51	0.38	0.71	4.37
Number per 1 min driving at speed limit	12.67	2.64	1.76	2.76	0.83	0.36	0.26	0.50	3.06

[a]Including speed limit signs. [b]Mostly associated with road construction; appearing as obvious to drivers, but not specified in the statute. [c]Subjective perception. Cf. discussion below related to intersections

Fig. 11.1 Concentration of road signs (14 signs at a distance of 85 m). Unnecessary repeat on the left side of road forbidding pedestrian entry (correctness rating 1), but lack of repeat of sign warning of bidirectional road (correctness rating 0)

In this preliminary analysis are given only a few examples of absurdities in road signage amongst a plethora observed. Firstly, quite interesting is a situation with signs that should be obvious to any driver but are not specified in the law and as such could be considered as illegal according to the letter of the statute [23]. Such signage is frequently not only allowed, but also proposed by the road administrator [24]; however, as an entity not permitted to establish or modify a statute, these signs cannot be considered as enforceable. This type of road signs was classified as unofficial signs; in majority of cases they were associated with road work signage. An example is given in Fig. 11.2: a construction zone road sign informing of road narrowing is clear, but it is not only unofficial, but also an example of poor quality (face bleached-out) and providing incorrect information (announcement of road narrowing 400 m ahead, while the narrowing starts just 70 m after the sign). Another example of incorrectness is shown in Fig. 11.3—a warning of a road deviation that was just passed; in addition erroneous horizontal markings, forcing all drivers to cross double solid line, is obvious.

Another example of absurd road signs is lowering of speed limit at a distance of only 30 m, shown in Fig. 11.4. While the road administrator could definitely produce a technical reason for such signage, it is incomprehensible to vast majority of drivers who simply ignore it. That sign is also wrong because, unlike in other similar areas where road was narrowed, there was no prior decrease of speed limit. In addition, the information sign with road lane arrangement that accompanies cancellation of the ultra-short 40 km/h zone is incorrect, because such arrangement starts about 120 m farther.

Table 11.4 Assessment of vertical road signs

	All	Warning	Prohibitory[a]	Speed limit	Mandatory	Information	Direction	Unofficial[b]	Other
Number	2022	421 (20.8%)	801 (39.6%)	441 (21.8%)	132 (6.5%)	57 (2.8%)	42 (2.1%)	80 (4.0%)	489 (24.2%)
Repeats on left side	572 (28.3%)	98 (23.3%)	275 (34.3%)	149 (33.8%)	16 (12.1%)	12 (21.1%)	2 (4.8%)	21 (26.3%)	148 (30.3%)
Non-repeat signs on the left	16 (0.8%)	3 (19%)	6 (37%)	3 (19%)	4 (25%)	–	–	–	3 (19%)
Average correctness rating	2.65	2.90	2.50	2.44	2.90	2.93	3.00	2.58	2.57
Incorrect (ratings 0-2)	644 (31.8%)	33 (7.8%)	371 (46.3%)	223 (50.6%)	13 (9.8%)	2 (3.5%)	–	25 (31.3%)	200 (40.9%)
Average visibility rating	2.94	2.94	2.95	2.91	3.00	3.00	3.00	2.50	2.98
Poor visibility (ratings 0-2)	70 (3.5%)	14 (3.3%)	24 (3.0%)	24 (5.4%)	–	–	–	24 (30.0%)	8 (1.6%)
Average quality rating	2.92	2.98	2.91	2.87	2.97	3.00	2.81	2.49	2.96
Inadequate quality (ratings 0-2)	114 (5.6%)	6 (1.4%)	53 (6.6%)	43 (9.8%)	4 (3.0%)	–	4 (9.5%)	25 (31.3%)	22 (4.5%)
Unnecessary[c]	728 (36.0%)	13 (3.1%)	329 (41.1%)	189 (42.9%)	8 (6.1%)	2 (3.5%)	0 (0%)	1 (1.3%)	186 (38.0%)
Erroneous	114 (5.6%)	8 (1.9%)	45 (5.6%)	26 (5.9%)	1 (0.8%)	0 (0%)	1 (2.4%)	2 (2.5%)	31 (6.3%)

[a]Including speed limit signs. [b]Mostly associated with road construction; appearing as obvious to drivers, but not specified in the statute. [c]Subjective perception. Cf. discussion below related to intersections

Fig. 11.2 Unofficial signage that should be clear to all drivers. However, the signs are not only incorrect (distance of 70 m is displayed as 400 m; correctness rating 1), but also weakly visible (quality rating 1), and signs are misaligned (visibility rating 1)

Fig. 11.3 Incorrect horizontal signage forcing drivers to cross double solid line and wrong vertical signage (correctness rating 1)

11.4.4 Road Markings

Because of the road work zone, yellow markings were considered primary. It appears that while there are legal requirements for quality of permanent marking [23], there are missing such demands for the temporary yellow signalisation. Overall, road marking quality and clarity were subjectively rated as very poor, particularly at the most dangerous places like intersections and roundabouts. Research of drivers' visual

Fig. 11.4 Speed limit 40 km/h, but zone only 30 m long and imposed too late (correctness rating 1), crooked information sign (visibility rating 2); followed by inaccurate information signs on right with a repeat on left (correctness rating 1)

perception confirmed the importance of horizontal road marking in such places [25]. In numerous cases, yellow horizontal markings formed a maze of more and less visible lines, a result of traffic rerouting and inappropriate coverage of obsolete signalisation. In a few cases, they were forcing drivers to break rules of the road, as shown in Fig. 11.3.

An example of inadequate signage (both horizontal and vertical) is shown in Fig. 11.5—besides severely worn yellow markings at the roundabout, the open exit ahead (with incorrectly positioned no entry sign—only on the left side, so according

Fig. 11.5 Inadequate and inappropriate signage—road straight ahead is only for construction vehicles. Yellow road markings almost completely worn

Fig. 11.6 Horizontal road markings directing in wrong direction

Fig. 11.7 White horizontal road markings in a roadwork zone and a construction access

to the letter of the law carrying to value) appears as the main road. Another example is given in Fig. 11.6, where the clearly visible yellow markings guide drivers directly into travel lane in opposite direction. In Fig. 11.7 it is shown that after the traffic was rerouted to a newly build carriageway, only white markings were used. This is both incorrect and inefficient, because the white markings do not appear to be positioned according to the final roadway configuration. Multiple additional examples could be given.

11.4.5 Other

At the test stretch there were 13 marked pedestrian crossings, 2 (15%) of them controlled with traffic lights. There were no pedestrians were using them; quality of 'zebra' horizontal markings was very poor, quite contrary to demands in developed countries that care about traffic and pedestrian safety [26].

Fig. 11.8 End of no parking zone that would be cancelled by intersection (correctness rating 2)

There were 8 roundabouts and 8 intersection controlled with traffic lights; in all of the cases the light was green, but it should be noted that some of these traffic lights were established solely to permit construction vehicles access. Assessment of intersections and pedestrian crossings is beyond the scope of this preliminary report while appropriate methodology is being developed.

There were 12 parked or stopped vehicles, including 6 lorries (other than construction vehicles with orange flashing lights). All of them were violating traffic rules; no orange warning triangles were displayed. An attention should be paid to an accident (a rear-ending): three people were walking on the roadway without safety vests, no warning triangle was displayed, and the traffic had to manoeuvre through cones and use travel lane designated for the opposite direction.

11.5 Discussion

There may be several reasons for disobedience of speed limits, from individual psychological traits, through lack of comprehension of own and vehicular limitations, adjustment of own speed to perceived as reasonable in particular environment, to the lack of knowledge about the set limit [27]. In case of the assessed road stretch, the experienced test driver's subjective opinion is that speed limit of 60 km/h is reasonable given the traffic load, lane width, highly inadequate horizontal signalisation, and extremely poor condition of roadway surface. However, the occasional limits to 40 km/h are subjectively perceived as unnecessary and contradicting the common sense of the drivers. After rerouting of the road to a newly built carriageway, after applying appropriate road markings, speed limit of 80 km/h would be perceived as appropriate.

Second discussion item is excessive number of road signs, such as shown in Fig. 11.1. If it is not possible by an average driver travelling at the posted speed limit to recognise all of the road signs, how could one expect the driver to obey them? If one adds such situations to other inconsistencies and inaccurate information provided by some signs (like shown in Fig. 11.2 or in Fig. 11.3), notorious violations of the traffic rules are becoming more easily understood.

Furthermore, one must consider whether appropriateness of road signs. According to rules of the road in Poland [28], an intersection is cancelling temporary limitations. An intersection is defined as level crossing or forking of roads other than crossing a driveway or a dirt road; dirt road is defined as a road without hard paving at a distance of 20 m. Hence, a doubt arises whether an access for construction vehicles can (should?) be treated as an intersection or not, particularly if it does not meet the requirement of 20 m paved surface. There were 96 of such accesses at the test stretch. Consequently, if such accesses are not intersections, a plethora of road signs (such as repeats of speed limits) were unnecessary because they were not cancelled. The subjective assumption in this chapter was based on a reasoning that if road signs indicated an intersection, it was an intersection even if it did not meet its legal requirements. An example of such construction access is given in Fig. 11.7.

As an example one must also consider situation shown in Fig. 9: was it a traffic lights-controlled construction access or an intersection? If it was an intersection, the road sign cancelling no stopping zone was unnecessary because the intersection would cancel it and it is not legal to stop within 10 m of an intersection (this was the assumption in considering necessity of road signs in Tables 11.3 and 11.4). On the other hand, if this were not an intersection then the end road sign ending the no stopping zone would be correct, but subsequent prohibitive road signs would be unnecessarily redundant. Should one argue that a superfluous road sign is not causing harm (or causing less harm than possible later misinterpretation), it must be reminded that the time allocated for observation of such sign was taken from observing the road ahead. The doubts that remain are an example of the lack of clarity and coherence of road signs in Poland; this is seen as one of the reasons for their notorious violation.

While doing such analysis, one should deliberate whether elsewhere driver behaviour would be similar. To compare the speeds of overtaking vehicles at road-work zones, a similar test drive was done in Czech Republic, where road safety record is much better in spite of imperfect infrastructure [29]. The question is whether speeding in Poland, being so common, can be correlated with the number of accidents. The details provided in Table 11.5 do not appear to require further comments. These enormous accidents costs, even without injuries, severely burden the national budget. The time lost and additional environmental pollution due to being delayed by a collision in such a construction zone is not included, even though these impacts could be very meaningful. Nonetheless, whereas there appears to be a direct correlation, one must not exclude other factors, like traffic load and type, roadway features, signage, etc.

Enforcement of speed limits and traffic rules in Poland is not only weak, but also very spotty and highly inconsistent. All stationary photo radars are marked with warning signs, so drivers slow down before them, only to accelerate again. Police speed traps are perceived as placed highly inappropriately; they are very frequently in the same locations, which are known by local drivers. During the test drive, only 2 marked police vehicles were seen, in both instances with a vehicle pulled over. If

Table 11.5 Speeds in work zones in Poland and in Czech Republic

	Poland	Czech Republic
AADT (in parentheses: percentage heavy vehicles)	33,880 (31%)	41,120 (28%)
Stretch length [km]	152.3	43.8
Speed limit (in parentheses speed of test vehicle) [km/h]	56 (58)	80 (72)
Overtaking vehicles per 1 km	2.46	0.14
Average speed of overtaking vehicles [km/h] (excess)	97 (33)	83 (0)
Maximum overtaking speed [km/h] (excess)	162 (96)	91 (4)
Number of road signs per 1 km (per intersection)	13 (18)	8 (10)
Number of accidents (per 100 km of the test stretch per month)	209	44
Cost of these accidents [PLN][a]	6,025,522	1,268,531

[a]Assumed is cost per collision without injuries of 28,830 PLN (172,224 CZK or 6,709 EUR or 7,509 USD), per official exchange rates, based on estimate of accidents costs for 2019 [KRBRD, 2019][2]

one assumes that one driver was pulled over and penalised for speeding, there could be 200 other, equally speeding, that were not noticed or ignored by the same police patrol. Consequently, it is difficult not to concur with the anecdotal evidence that in Poland one who receives a traffic violation penalty feels that it was not because of violating rules of the road, but solely because of being unlucky.

11.6 Conclusions

A new, efficient, and simple methodology for naturalistic driving study that excludes the test driver, but instead concentrates on behaviour of other road users and road features was described and proof of its feasibility was shown. The methodology is planned to be applied for assessment of other road stretches in Poland and its expansion to another road features is also being explored. It is hoped that not only the procedure, but also outcome would be utilised by road administrators to make simple and inexpensive road improvements that can meaningfully increase road safety. Based on this preliminary study, validity of the proposed methodology is confirmed.

The results from assessment of a long road construction zone clearly indicate (1) excessive speeds, (2) overabundant number or road signs, (3) their low correctness, and (3) highly inadequate road marking. Hence, it is quite possible that speeding and violation of traffic rules in Poland is associated not only with personal traits of the drivers and the complete lack of 'safety climate' [30], but also with inappropriate road signage and the absence of self-explaining road features [31].

Any driver at the analysed road stretch (and everywhere elsewhere in Poland) is attacked with constantly changing speed limits, very frequently seemingly placed in random locations and without any reason; consequently, speed limits are perceived as wrong and thus ignored. Profuse quantity of road signs, many with them incorrect and placed with mediocre attention, add to the perception of chaos and take drivers' attention from road ahead to their observation. Furthermore, horizontal road markings at this stretch only confirm the confusion. Given these, numerous driving violations could be treated as driving errors caused by inadequate infrastructure and unclear signage; it is quite possible that many drivers ignore the signs and rules as making no practical sense. This reasoning could be verified through appropriately designed survey, which work is planned in the near future.

Limitations of the used methodology include the use of only one driver, with driving habits based on experience. In addition, the influence of momentary circumstances may have affected the outcome. Error analysis and statistical validation were beyond the scope of this preliminary report.

References

1. Police Statistics. https://statystyka.policja.pl/st/ruch-drogowy/76562,Wypadki-drogowe-raporty-roczne.html. Accessed 22 Oct 2020
2. Krajowa Rada Bezpieczeństwa Ruchu Drogowego (KRBRD) Wycena kosztów wypadków i kolizji drogowych na sieci dróg w Polsce na koniec roku 2018, z wyodrębnieniem średnich kosztów społeczno-ekonomicznych wypadków na transeuropejskiej sieci transportowej. Warszawa, 2019. https://www.krbrd.gov.pl/pl/koszty-zdarzen-drogowych.html. Accessed 21 Oct 2020
3. Giuffrè O, Granà A, Tumminello ML, Giuffrè T, Trubia S, Sferlazza A, Rencelj M (2018) Evaluation of roundabout safety performance through surrogate safety measures from microsimulation. J Adv Transp, 1–14. Article ID 4915970
4. Lynnyk I, Kulbashna N, Galkin A, Prasolenko O, Dulfan S (2020) Safety assessment of adjacent roads sections via maximum entropy driver's perception field. Communications-Sci Lett Univ Zilina 22(4):182–190
5. Macioszek E, Sierpiński G, Staniek M (2017) Analysis of trends in development of freight transport logistics using the example of Silesian Province (Poland)—a case study. Transp Res Proc 27:388–395
6. Macioszek E, Sierpiński G (2020) Charging stations for electric vehicles—current situation in Poland. In: Mikulski J (ed) Research and the future of telematics, vol 1289. Commun Comput Inf Sci. Springer, Heidelberg, pp 124–137
7. Wijayaratna KP, Cunningham ML, Regan MA, Jian S, Chand S, Dixit VV (2019) Mobile phone conversation distraction: understanding differences in impact between simulator and naturalistic driving studies. Accident Anal Prevent 129:108–118
8. Barnard Y, Utesch F, van Nes N, Eenink R, Baumann M (2016) The study design of UDRIVE: the naturalistic driving study across Europe for cars, trucks and scooters. Eur Transp Res Rev 8:14. https://doi.org/10.1007/s12544-016-0202-z
9. Özkan T, Lajunen T, Summala H (2006) Driver behaviour questionnaire a follow-up study. Accident Anal Prevent 38:386–395
10. Tay R, Knowles D (2004) Driver inattention: drivers' perception of risks and compensating behaviours. IATSS Res 28(1):89–94
11. Sundfør HB, Sagberg F, Høye A (2019) Inattention and distraction in fatal road crashes—results from in-depth crash investigations in Norway. Accident Anal Prevent 125:152–157
12. Noy IY, Shinar D, Horrey WJ (2018) Automated driving: safety blind spots. Safety Sci 102:68–78
13. Hauer E (2019) On the relationship between road safety research and the practice of road design and operation. Accident Anal Prevent 128:114–131

14. Kar K, Blankenship MR (2010) Road safety audit: findings from successful applications in Arizona. Transp Res Record J Transp Res Board 2182:113–120
15. Generalna Dyrekcja Dróg Krajowych i Autostrad: Generalny Pomiar Ruchu 2015, Średni Dobowy Ruch Roczny (SDRR) W punktach pomiarowych w 2015 roku na Drogach Krajowych, https://www.gddkia.gov.pl/userfiles/articles/g/generalny-pomiar-ruchu-w-2015_15598//SYNTEZA/WYNIKI_GPR2015_DK.pdf
16. Fredette M, Mambu LS, Chouinard A, Bellavance F (2008) Safety impacts due to the incompatibility of SUVs, minivans, and pickup trucks in two-vehicle collisions. Accident Anal Prevent 40(6):1987–1995
17. Keall MD, Newstead S (2008) Are SUVs dangerous vehicles? Accident Anal Prevent 40:954–963
18. Murphy P, Morris A (2020) Quantifying accident risk and severity due to speed from the reaction point to the critical conflict in fatal motorcycle accidents. Accident Anal Prevent 141:
19. Burghardt TE, Mosböck H, Pashkevich A, Fiolić M (2020) Horizontal road markings for human and machine vision. Transp Res Proc 48:3622–3633
20. Leviäkangs P (1998) Accident risk of foreign drivers—the case of Russian drivers in South-Eastern Finland. Accident Anal Prevent 30(2):245–254
21. Hughes PK, Cole BL (1986) What attracts attention when driving? Ergonomics 29(3):377–391
22. Costa M, Simone A, Vignali V, Lantieri C, Bucchi A, Dondi G (2014) Looking behavior for vertical road signs. Transp Res Part F 23:147–155
23. Rozporządzenie Ministra Infrastruktury z dnia 3 lipca 2003 r. w sprawie szczegółowych warunków technicznych dla znaków i sygnałów drogowych oraz urządzeń bezpieczeństwa ruchu drogowego i warunków ich umieszczania na drogach (Dz.U. 2003 nr 220 poz. 2181). Załącznik 2. Szczegółowe warunki techniczne dla znaków drogowych poziomych i warunki ich umieszczania na drogach, http://isap.sejm.gov.pl/isap.nsf/DocDetails.xsp?id=WDU20032202181&type=2
24. Generalnego Dyrektora Dróg Krajowych i Autostrad (GDDKiA): Zarządznie nr 34 z dnia 30 lipca 2014 roku w sprawie typowych schematów oznakowania robót oraz pomiarów diagnostycznych prowadzonych w pasie drogowym. Załącznik 1. Katalog typowych schematów oznakowania robót prowadzonych w pasie drogowym, https://www.gddkia.gov.pl/userfiles/articles/z/zarzadzenia-generalnego-dyrektor_13901/zarzadzenie%2020.pdf
25. Pashkevich A, Burghardt TE, Shubenkova K, Makarova I (2020) Analysis of drivers' eye movements to observe horizontal road markings ahead of intersections. In: Varhelyi A, Žuraulis V, Prentkovskis O (eds) Vision zero for sustainable road safety in Baltic Sea Region. VISZERO 2018. Lecture Notes in Intelligent Transportation and Infrastructure, pp 1–10. Springer, Cham, https://doi.org/10.1007/978-3-030-22375-5_1
26. Burghardt TE, Pashkevich A, Mosböck H (2019) Yellow pedestrian crossings: from innovative technology for glass beads to a new retroreflectivity regulation. Case Stud Transp Policy 7(4):862–870
27. Kanellaidis G, Golias J, Zarifopoulos K (1995) A survey of drivers' attitudes toward speed limit violations. J Safety Res 26(1):31–40
28. Obwieszczenie Marszałka Sejmu Rzeczypospolitej Polskiej z dnia 9 grudnia 2019 r. w sprawie ogłoszenia jednolitego tekstu ustawy—Prawo o ruchu drogowym (Dz.U. 2020 poz. 110), https://isap.sejm.gov.pl/isap.nsf/DocDetails.xsp?id=WDU20200000110
29. Ambros J, Turek R, Brich M, Kubeček J (2019) Safety assessment of Czech motorways and national roads. Eur Transp Res Rev 11:1. https://doi.org/10.1186/s12544-018-0328-2
30. Neal A, Griffin MA (2006) A study of the lagged relationships among safety climate, safety motivation, safety behavior, and accidents at the individual and group levels. J Appl Psychol 91(4):946–953
31. Gitelman V, Pesahov F, Carmel R, Bekhor S (2016) The identification of infrastructure characteristics influencing travel speeds on single-carriageway roads to promote self-explaining roads. Transp Res Proc 14:4160–4169

Chapter 12
Methods of Parking Measurements—Research of Parking Characteristics in Paid Parking Zones with Dynamic Parking Information

Agata Kurek

12.1 Introduction

Road traffic in the city is a complex process. It is characterized by differentiation in the existing transport subsystems, between city areas or outside of it, as well as in particular periods of the day, week, month, and year [1, 2]. Measurements of road traffic parameters are an important element of traffic engineering. They make it possible to learn about the communication behavior of road users, the impact of the occurrence of individual transport subsystems on traffic conditions, changes in the city's transport system, the impact of transport on the environment [3–6]. Performing road traffic measurements provides data for the analysis of road traffic conditions for scientists looking for new solutions to improve road traffic conditions and road safety. Moreover, they are carried out to properly manage road traffic in the city by transport planners and road managers. Technological progress made it possible to improve the performance of road traffic measurements. This reduced the time needed to collect the necessary data and reduced the likelihood of human error.

The final stage of any car journey is to park the vehicle. The increase in traffic in the city causes not only a longer journey time, a lower level of road safety or the negative impact of transport on the environment but also the problem of finding a space for the driver to leave the vehicle. In the areas of Polish cities with a shortage of parking spaces, it is possible to introduce a Paid Parking Zone (PPZ). Searching for space by drivers to leave the vehicle may contribute to increased traffic near the parking due to the so-called search traffic. The purpose of introducing Dynamic Parking Information (DPI) is to guide drivers to empty parking spaces. This contributes to the reduction of the time needed by the driver to find an empty parking space.

A. Kurek (✉)
Faculty of Transport and Aviation Engineering, Department of Transport Systems and Traffic Engineering, Silesian University of Technology, Gliwice, Poland
e-mail: Agata.Kurek@polsl.pl

E. Macioszek and G. Sierpiński (eds.), *Research Methods in Modern Urban Transportation Systems and Networks*, Lecture Notes in Networks and Systems 207, https://doi.org/10.1007/978-3-030-71708-7_12

Table 12.1 Possibility of determining parking characteristics measures with the use of individual measuring techniques

Parking characteristics measures	Measurement techniques		
	Patrol research	Motion detectors	Video cameras
Parking spaces usage rate	✓	✓	✓
Rotation indicator	✓	✓ (Depending on the data collecting system, it can be accurate to the parking space or the parking area)	✓
Parking time	✓	✓ (depending on the data collecting system, it can be accurate to the parking space or the parking area)	✓
Accumulation	✓	✓	✓
Parking intensity	✓	✓	✓
Parking filling indicator in peak	✓	✓	✓
Share of vehicles, which parking inconsistently with the regulations	✓	–	✓
Search time	–	–	✓

where: ✓-possible; – - impossible

The measurement method depends on the type of car park and the characteristics to be tested. Knowledge of the measures of parking characteristics in a given area allows for better management of the parking space. Parking surveys make it easier for city authorities to decide on the introduction of PPZ and to determine the parking rate. It is possible to verify whether the parking cost is appropriate in the case of measurements on the already existing PPZ. Carrying out analyzes of the occupancy of parking spaces and the rotation indicator provides information on whether parking spaces located in one area are characterized by differentiation or not. The introduction of the DPI is justified in the case of large differences in the values of the occupancy of the parking space and the rotation indicator. DPI can contribute to more efficient use of parking space. Measurements of parking characteristics to determine the time a driver searches for a parking space may also be the basis for introducing a DPI.

The measures of parking characteristics that are used for the analysis include [7]:

- parking space usage rate—the ratio of the number of vehicles to the number of parking spaces in the analyzed area in a given analysis period [%],
- turnover rate—the average number of parking vehicles using one parking space in the analyzed period [P/parking space],
- parking time—length of vehicle parking [s],
- accumulation—the number of vehicles parked simultaneously in the analyzed area [P],

- parking intensity—number of vehicle-hours of parking in the analyzed parking at a given time of the day [Ph],
- parking filling at peak—the ratio of the peak number of vehicles parking simultaneously in the analyzed period to the number of parking spaces in the analyzed area [%],
- share of vehicles, which parking inconsistently with the regulations - ratio of the number of vehicles parked contrary to the regulations to the number of all vehicles parked in the analyzed area [-],
- search time—the time the driver looks for a parking space [s].

The chapter aims to present the methods that are used to carry out parking measurements. After the introduction, the second subchapter presents an overview of the work on measuring methods in parking. The next subchapter presents the results of the study of parking characteristics in the area covered by the Paid Parking Zone and Dynamic Parking Information.

12.2 Methods of Parking Measurements

In the literature on the subject, works on the study of road traffic characteristics present various measurement techniques. These are mainly data collected using:

- induction loops: Coifrman [8], R. Chrobok et al. [9],
- video cameras: Zafri et al. [10], Tsuji et al. [11], drones Barmpounakis and Geroliminis [12], Kaaniche et al. [13],
- ANPR cameras: Zheng [14],
- smartphones Mohan [15], Mednis et al. [16], Singh et al. [17], etc.

In parking, researches are carried out to learn about the characteristics of parking, analyze the quality of measurements carried out with various methods and look for solutions aimed at increasing the availability of parking spaces, as well as reducing the time needed to search for a parking space, and thus reducing the impact of parking to traffic conditions.

One of the methods of carrying out measurements in parking is a patrol survey. Due to technological progress, measurements are made in this way less and less. The authors of Cao and Menendez [18] analyzed the relationship between the accuracy of the average parking time measurem[illegible] the intensity of the patrol survey. The paper [19] presents an analysis of parking characteristics based on data obtained from a patrol survey in Dhaka. The research results show that a large proportion of drivers leave their vehicle for more than 3 h, which results in low rotation indicator values.

Another method of carrying out parking measurements is research with the use of video cameras. Authors Cai et al. [20], De Almeida et al. [21], Huang and Wang [22], Bibi et al. [23], Basavaraju [24] presented a method of research on parking with the use of cameras, which based on automatic image analysis allows determining whether a given parking space is occupied. The work [25] also presents the use of video cameras to analyze the occupancy of parking spaces. Additionally, this study presents the possibility of recognizing the license plates that are stored. Such data can be used to analyze the share of vehicles registered outside the city in which the measurements are made to the number of vehicles registered in a given city. In turn, the authors R. C. Hampshire et al. [26] used the measurement method with the use of video cameras to analyze the time a driver searches for an empty parking space.

Automatic data collection from motion detectors can be used to carry out measurements in parking. The paper [27] presents the possibility of using Wireless Sensor Networks (WSN) to detect vehicles in parking spaces. The authors indicate that this solution can be used to inform drivers about empty parking spaces. The use of WSN for the analysis of the occupancy of parking spaces was also presented by the authors, Karbab et al. [28], Patil, and Bhonge [29]. WSN and radio frequency identification (RFID) were used to analyze empty parking spaces and indicate them to drivers looking for a place to leave the vehicle. Other research in this area are related to the electric vehicles [30] or freight transport [31].

Table 12.1 shows the parking characteristics that can be determined after researching with the use of particular measurement methods. Measurements with the use of video cameras provide the most data. However, they require a lot of time and effort to process and analyze the data. This is due to the fact that in the case of researches with the use of a video camera, a person is required to supervise the measurement process, analyze the collected material, and determine the parking characteristics. The publicly available monitoring recordings on the Internet can be used to minimize the time needed to collect recordings. The screen recording program can be used to record images from these cameras. Therefore, research using the online camera is a less time-consuming measuring technique than research using the video camera.

12.3 Parking Characteristics Research at PPZ and DPI in Gliwice

For the chapter, data was obtained from the Municipal Roads Authority in Gliwice. The data concerned the number of vehicles parked in parking spaces covered by the Paid Parking Zone and Dynamic Parking Information (Fig. 12.1). In the analyzed area of PPZ and DPI, parking spaces are equipped with a magnetic sensor that detects the presence of the vehicle by disturbing the magnetic field generated by the induction loop. Data from sensors are collected by concentrators, from which data on the occupancy of parking spaces are sent to the controller. Information about empty parking spaces is displayed on a variable message table.

Figures 12.2, 12.3, 12.4 and 12.5 show the parking characteristics measures. The data presented in the graphs allow stating that the lowest values of rotation and occupancy of parking spaces occur in the morning and night hours 00:00–06:00 and

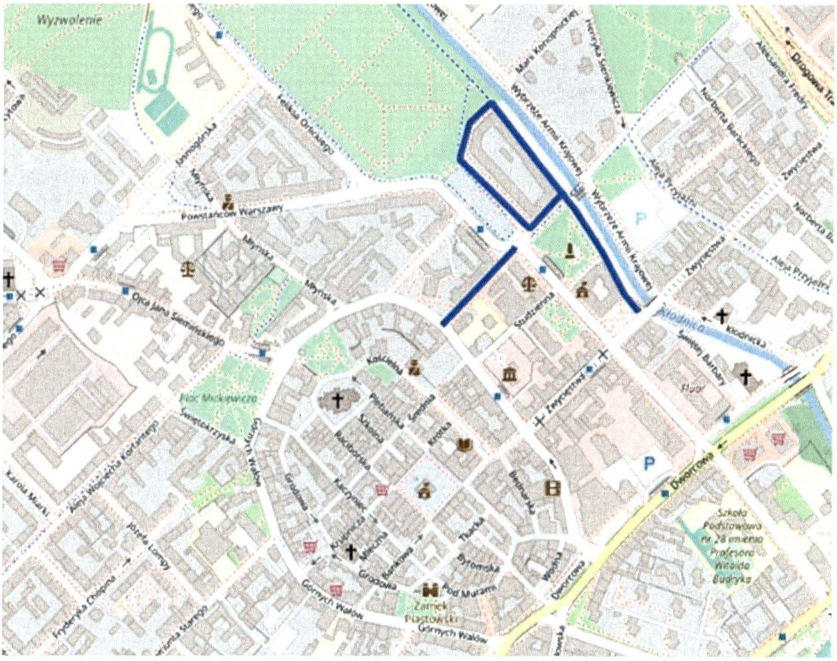

Fig. 12.1 The analyzed area in the Paid Parking Zone and Dynamic Parking Information in Gliwice

Fig. 12.2 Rotation on the analyzed parking places in particular periods of the day

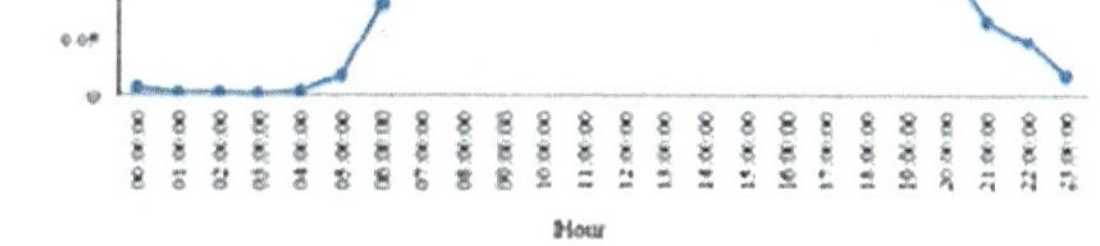

21:00–23:00. The rotation indicator increases from 07:00. Then there is a decrease and fluctuations in this value until 15:00 and then it decreases again. On the other hand, the occupancy rate in parking spaces increases from 08:00 and remains at a similar level until 15:00. The rotation indicator is the lowest on Saturdays and Sundays. It may be caused by not collecting fees for parking in the analyzed parking spaces on weekend days. On working days of the week, the value of the rotation indicator is at a similar level, it is the highest on Friday. However, in the case of the occupancy of parking spaces on weekend days, a decrease in value is also observed,

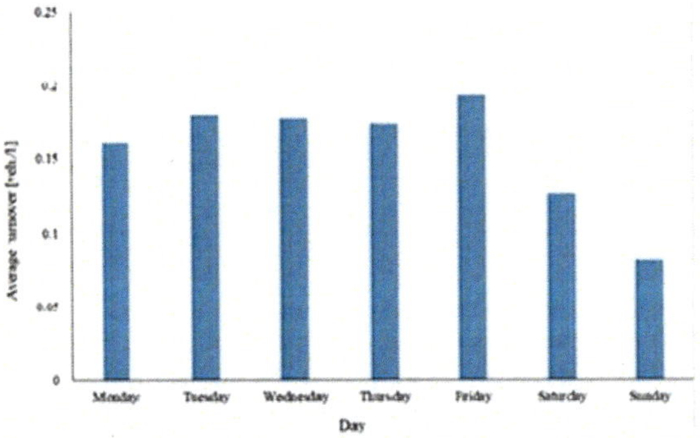

Fig. 12.3 Rotation of the analyzed parking spaces on particular days of the week

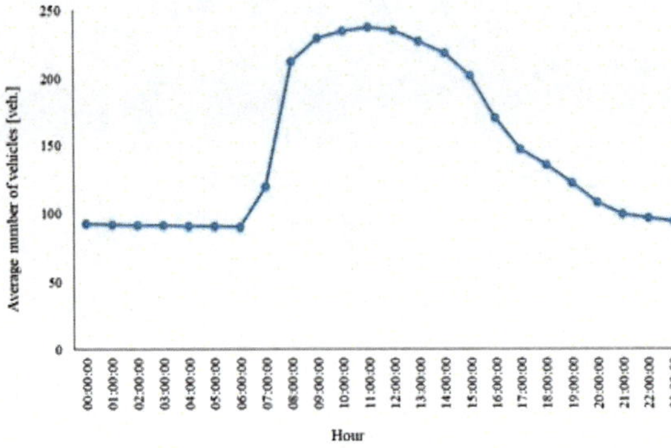

Fig. 12.4 Occupancy in parking spaces in the analyzed area in particular periods of the day

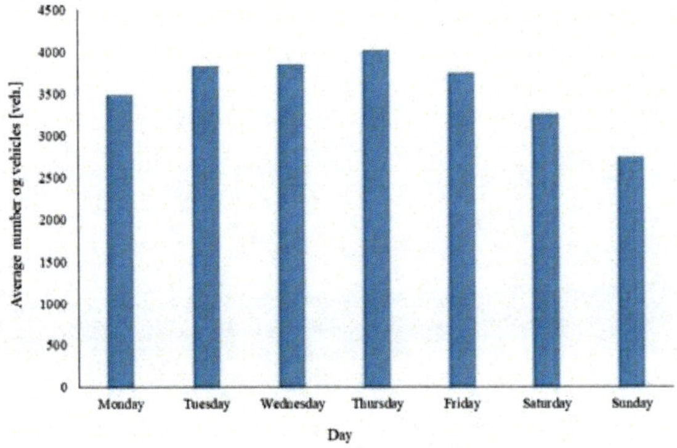

Fig. 12.5 Occupancy in parking spaces in the analyzed area on particular days of the week

but it is smaller than in the case of the rotation indicator. On working days, the value of the rotation indicator is at a similar level, it is the highest on Thursdays.

In turn, Fig. 12.6 shows the length of vehicle parking in parking spaces in the analyzed area. The people leave their vehicle for a period of 10 min to 2 h and 8–24 h. The short length of parking in parking spaces may indicate that traveling aims to arrange short errands in the city center (visit to the city office, shopping, medical

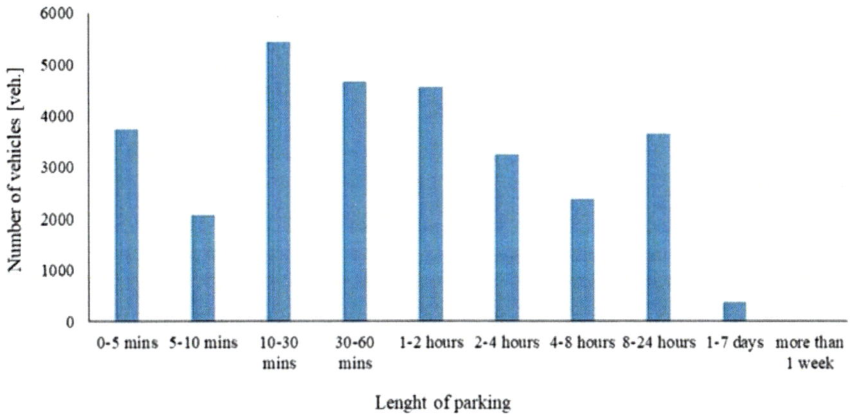

Fig. 12.6 Length of parking in parking spaces in the analyzed area

care, etc.). However, a parking time of 8–24 h may result from leaving the vehicle in these parking spaces by residents of the buildings located near the analyzed area.

12.4 Conclusions

The chapter aimed to present the methods that are used to conduct parking measurements. Technological progress allows for the search for new solutions aimed at facilitating measurements in parking. It contributes to the reduction of the time needed to collect relevant data that allow for conducting analyzes. Conducting this type of research is necessary due to the increasing share of individual transport in travel.

Based on the data collection method, basic parking characteristics were determined automatically from magnetic sensors. The data included parking spaces in the Paid Parking Zone and Dynamic Parking Information in Gliwice. The analysis of parking characteristics allows for the following conclusions:

- the rotation indicator was characterized by the highest values at 06:00 in the morning peak and 15:00 in the afternoon peak. In the period between the peaks, the value of the rotation indicator was at a similar level. On the other hand, the lowest values of the rotation indicator (close to 0) can be observed between 20:00–23:00 and 00: 00–06: 00 and on weekend days,
- parking fees in the analyzed PPZ are collected on working days between 09:00 and 17:00. It can therefore be concluded that PPZ does its task, i.e. increases the availability of parking spaces for a larger number of drivers,
- the highest value of the occupancy of parking spaces took place between 07:00 and 15:00 and on days from Monday to Friday. On the other hand, the lowest values occurred between 20:00–23:00 and 00:00–06:00 and on weekend days—Saturdays and Sundays,

- most drivers leave their vehicles in the analyzed area for a period from 10 min to 2 h and 8–24 h,
- the length of parking in parking spaces may indicate that traveling aims to arrange short errands in the city center (in the case of a 10-minute-2 h parking). However, a parking time of 8–24 h may result from leaving the vehicle in these parking spaces by residents of the buildings located near the analyzed area.

Acknowledgements The present research has been financed from the "Excellence Initiative—Research University" program implemented at the Silesian University of Technology, 2020-2022 as a part of a grant entitled "Analysis of parking characteristics in the conditions of PPZ and DPI functioning in selected areas of GZM cities".

References

1. Tubis A (2018) Risk assessment in road transport—strategic and business approach. J KONBiN 45(1):305–324
2. Macioszek E, Świerk P, Kurek A (2020) The bike-sharing system as an element of enhancing sustainable mobility—a case study based on a city in Poland. Sustainability 12:1–29
3. Macioszek E (2018) The comparison of models for follow-up headway at roundabouts. In: Macioszek E, Sierpiński G (eds.) Recent advances in traffic engineering for transport networks and systems. Lecture Notes in Networks and Systems 21. Springer International Publishing, Switzerland, pp 16–26
4. Szczuraszek T, Macioszek E (2013) Proportion of vehicles moving freely depending on traffic volume and proportion of trucks and buses. Baltic J Road Bridge Eng 8:133–141
5. Macioszek E (2012) Geometrical Determinants of Car Equivalents for Heavy Vehicles Crossing Circular Intersections. [w:] J. Mikulski (ed.) Transport Systems Telematics. Communications in Computer and Information Science329. Springer-Verlag, Berlin, Heidelberg, pp 221–228
6. Macioszek E, Sierpiński G, Czapkowski L (2010) Problems and issues with running the cycle traffic through the roundabouts. In: [w:] Mikulski J (ed) Transport systems telematics. communications in computer and information science 104. Selected Papers. Springer-Verlag, Berlin, Heidelberg 2010, pp 107–114
7. Gaca S, Suchorzewski W, Tracz M (2009) Inżynieria ruchu drogowego: teoria i praktyka. Wydawnictwo Komunikacji i Łączności
8. Coifman B (2002) Estimating travel times and vehicle trajectories on freeways using dual loop detectors. Transp Res Part A Policy Practice 36:351–364
9. Chrobok R, Kaumann O, Wahle J, Schreckenberg M (2004) Different methods of traffic forecast based on real data. Eur J Oper Res 155:558–568
10. Zafri NM, Rony AI, Adri N (2019) Analysis of pedestrian crossing speed and waiting time at intersections in Dhaka. Infrastructures 4:1–20
11. Tsuji T, Hattori H, Watanabe M, Nagaoka N (2002) Development of night-vision system. IEEE Trans Intell Transp Syst 3:203–209
12. Barmpounakis E, Geroliminis N (2020) On the new era of urban traffic monitoring with massive drone data: The pNEUMA large-scale field experiment. Transp Res Part C Emerg Technol 111:50–71
13. Kaaniche K, Champion B, Pégard C, Vasseur P (2005) A vision algorithm for dynamic detection of moving vehicles with a UAV. In: Proceedings of the 2005 IEEE international conference on robotics and automation, pp 1878–1883. IEEE
14. Zheng F, Li J, Van Zuylen H, Liu X, Yang H (2018) Urban travel time reliability at different traffic conditions. J Intell Transp Syst 22:106–120

15. Mohan P, Padmanabhan VN, Ramjee R (2008) Nericell: rich monitoring of road and traffic conditions using mobile smartphones. In: Proceedings of the 6th ACM conference on Embedded network sensor systems, pp 323–336
16. Mednis A, Strazdins G, Zviedris R, Kanonirs G, Selavo L (2011) Real time pothole detection using android smartphones with accelerometers. In: 2011 International conference on distributed computing in sensor systems and workshops (DCOSS), pp 1–6
17. Singh P, Juneja N, Kapoor S (2013) Using mobile phone sensors to detect driving behavior. In: Proceedings of the 3rd ACM symposium on computing for development, pp 1–2
18. Cao J, Menendez M (2012) Accuracy study of parking duration data from patrol survey. In: 12th Swiss Transport Research Conference (STRC 2012), pp 1–17
19. Zannat KE, Ahmed T, Mitra SK, Rafiq R, Hasan MA, Akhter K, Fahad ZH (2013) Parking demand and supply analysis of major shopping centers in Dhaka—a case study of new market shopping center along mirpur road. J Bangladesh Inst Planners 6:161–172
20. Cai BY, Alvarez R, Sit M, Duarte F, Ratti C (2019) Deep learning-based video system for accurate and real-time parking measurement. IEEE Internet Things J 6:7693–7701
21. De Almeida PR, Oliveira LS, Britto AS Jr, Silva EJ Jr, Koerich AL (2015) PKLot–A robust dataset for parking lot classification. Expert Syst Appl 42:4937–4949
22. Huang CC, Wang SJ (2010) A hierarchical bayesian generation framework for vacant parking space detection. IEEE Trans Circ Syst Video Technol 20:1770–1785
23. Bibi N, Majid MN, Dawood H, Guo P (2017) Automatic parking space detection system. In: 2017 2nd international conference on multimedia and image processing (ICMIP), pp 11–15
24. BS R (2015) Automatic smart parking system using Internet of Things (IOT). Int J Sci Res Publ 5:629–632
25. Dalarmelina NDV, Teixeira MA, Meneguette RI (2020) A real-time automatic plate recognition system based on optical character recognition and wireless sensor networks for ITS. Sensors 20:1–13
26. Hampshire RC, Jordon D, Akinbola O, Richardson K, Weinberger R, Millard-Ball A, Karlin-Resnik J (2016) Analysis of parking search behavior with video from naturalistic driving. Transp Res Record 2543:152–158
27. Yang J, Portilla J, Riesgo T (2012) Smart parking service based on wireless sensor networks. In: IECON 2012–38th Annual Conference on IEEE Industrial Electronics Society, pp. 6029–6034
28. Karbab E, Djenouri D, Boulkaboul S, Bagula A (2015) Car park management with networked wireless sensors and active RFID. In: 2015 IEEE international conference on electro/information technology (EIT), pp 373–378
29. Patil M, Bhonge VN (2013) Wireless sensor network and RFID for smart parking system. Int J Emerg Technol Adv Eng 3:188–192
30. Macioszek E, Sierpiński G (2020) Charging stations for electric vehicles—current situation in Poland. In: Mikulski J (ed) Research and the future of telematics, vol 1289. Communications in Computer and Information Science. Springer, Heidelberg, pp 124–137
31. Macioszek E, Sierpiński G, Staniek M (2017) Analysis of trends in development of freight transport logistics using the example of Silesian Province (Poland) - a case study. Transp Res Proc 27:388–395